This Book Belongs To

_____

_____

_____

1 How long does it take for 920 C to pass a given point in a wire that carries an electric current of 7.3 A?

2 Find the resistance of a resistor through which 620 C flow in 25 minutes if the potential difference across it is 4 V.

3 Find the electric resistance of a nickel rod 10 m long, if its cross–sectional area is 1.4 mm^2 and its resistivity is 6.53×10^{-8} Ω·m.

4 Calculate the resistivity of a wire 38 m long, if its cross–sectional area is 0.4 mm^2 and its electric resistance is 5.26 Ω.

5 Calculate the electric resistance of an aluminum wire 2.8 mm in diameter and 21 m long if its resistivity is 2.85×10^{-8} Ω·m.

6 A battery with an emf (electromotive force) of 5 V delivers 71.63 mA to a 65 Ω load.
   a) What is the internal resistance of the battery?
   b) What is its terminal potential difference when joined to this load?

7 An 87 μF capacitor is charged to a potential difference of 177 V. Find the charge on the capacitor.

8 The charge on a 99 μF capacitor is 10.2 mC. Find the potential difference to which the capacitor was charged.

9 When a capacitor is connected across an 139 V supply the charge is 65.33 mC. Determine its capacitance.

10 A certain light bulb has a resistance of 135 Ω. If a current of 1.56 A is going through it, calculate the voltage applied.

11 A voltage of 250 V is applied to an 150 Ω resistor. Find the current through the resistor.

12 A voltage of 100 V is applied to a resistor and the current going through is found to be 1.43 A. Find the resistance of the resistor.

13 A battery with an emf (electromotive force) of 6 V has internal resistance of 0.7 Ω. Find the voltage at the terminals of the battery when 78.2 mA is being drawn by an external load.

14 A steady electric current of 1.1 A flows through a wire. How many coulombs of charge pass through the wire per minute?

15 A charge of 2020 C was passed through a wire in 19 minutes. Calculate the average electric current during that interval.

16 How long does it take for 2520 C to pass a given point in a wire that carries an electric current of 4.7 A?

17 Find the resistance of a resistor through which 1120 C flow in 11 minutes if the potential difference across it is 3 V.

18 Find the electric resistance of a nichrome rod 43 m long, if its cross–sectional area is 0.3 mm^2 and its resistivity is 1.21×10^{-6} Ω·m.

19 Calculate the resistivity of a wire 43 m long, if its cross–sectional area is 2.3 mm^2 and its electric resistance is 1.98 Ω.

20 Calculate the electric resistance of a tin wire 0.5 mm in diameter and 21 m long if its resistivity is 1.18×10^{-7} Ω·m.

21 A battery with an emf (electromotive force) of 12 V delivers 292 mA to a 36 Ω load.
   a) What is the internal resistance of the battery?
   b) What is its terminal potential difference when joined to this load?

22 A 93 µF capacitor is charged to a potential difference of 3 V. Find the charge on the capacitor.

23 The charge on a 96 µF capacitor is 6.72 mC. Find the potential difference to which the capacitor was charged.

24 When a capacitor is connected across a 58 V supply the charge is 24.07 mC. Determine its capacitance.

25 A certain light bulb has a resistance of 185 Ω. If a current of 0.703 A is going through it, calculate the voltage applied.

26 A voltage of 160 V is applied to an 140 Ω resistor. Find the current through the resistor.

27 A voltage of 170 V is applied to a resistor and the current going through is found to be 0.791 A. Find the resistance of the resistor.

28 A battery with an emf (electromotive force) of 6 V has internal resistance of 1.7 Ω. Find the voltage at the terminals of the battery when 95.7 mA is being drawn by an external load.

29 A steady electric current of 5.4 A flows through a wire. How many coulombs of charge pass through the wire per minute?

30 A charge of 20 C was passed through a wire in 7 minutes. Calculate the average electric current during that interval.

31 How long does it take for 220 C to pass a given point in a wire that carries an electric current of 4.1 A?

32 Find the resistance of a resistor through which 1320 C flow in 9 minutes if the potential difference across it is 12 V.

33 Find the electric resistance of a copper rod 44 m long, if its cross–sectional area is 0.6 mm^2 and its resistivity is 1.67×10^{-8} Ω·m.

34 Calculate the resistivity of a wire 40 m long, if its cross–sectional area is 1.2 mm^2 and its electric resistance is 1.88 Ω.

35 Calculate the electric resistance of a gold wire 0.5 mm in diameter and 48 m long if its resistivity is 2.26×10^{-8} Ω·m.

36 A battery with an emf (electromotive force) of 10 V delivers 386.1 mA to a 23 Ω load.
    a) What is the internal resistance of the battery?
    b) What is its terminal potential difference when joined to this load?

37 A 40 μF capacitor is charged to a potential difference of 176 V. Find the charge on the capacitor.

38 The charge on a 35 μF capacitor is 665 μC. Find the potential difference to which the capacitor was charged.

39 When a capacitor is connected across an 121 V supply the charge is 48.4 mC. Determine its capacitance.

40 A certain light bulb has a resistance of 210 Ω. If a current of 0.524 A is going through it, calculate the voltage applied.

41 A voltage of 130 V is applied to an 105 Ω resistor. Find the current through the resistor.

42 A voltage of 160 V is applied to a resistor and the current going through is found to be 2.13 A. Find the resistance of the resistor.

43 A battery with an emf (electromotive force) of 4 V has internal resistance of 5.5 Ω. Find the voltage at the terminals of the battery when 61.1 mA is being drawn by an external load.

44 A steady electric current of 1.7 A flows through a wire. How many coulombs of charge pass through the wire per minute?

45 A charge of 1020 C was passed through a wire in 20 minutes. Calculate the average electric current during that interval.

46 How long does it take for 420 C to pass a given point in a wire that carries an electric current of 3.9 A?

47 Find the resistance of a resistor through which 720 C flow in 15 minutes if the potential difference across it is 22 V.

48 Find the electric resistance of a gold rod 21 m long, if its cross–sectional area is 1.1 mm^2 and its resistivity is 2.33×10^{-8} Ω·m.

49 Calculate the resistivity of a wire 25 m long, if its cross–sectional area is 2.4 mm^2 and its electric resistance is 0.181 Ω.

50 Calculate the electric resistance of a nichrome wire 1.6 mm in diameter and 34 m long if its resistivity is 1.25×10^{-6} Ω·m.

51 A battery with an emf (electromotive force) of 22 V delivers 766.6 mA to a 25 Ω load.
   a) What is the internal resistance of the battery?
   b) What is its terminal potential difference when joined to this load?

52 A 57 µF capacitor is charged to a potential difference of 180 V. Find the charge on the capacitor.

53 The charge on a 5 µF capacitor is 220 µC. Find the potential difference to which the capacitor was charged.

54 When a capacitor is connected across an 103 V supply the charge is 21.11 mC. Determine its capacitance.

55 A certain light bulb has a resistance of 170 Ω. If a current of 1.41 A is going through it, calculate the voltage applied.

56 A voltage of 200 V is applied to a 225 Ω resistor. Find the current through the resistor.

57 A voltage of 190 V is applied to a resistor and the current going through is found to be 1.12 A. Find the resistance of the resistor.

58 A battery with an emf (electromotive force) of 23 V has internal resistance of 5.2 Ω. Find the voltage at the terminals of the battery when 332 mA is being drawn by an external load.

59 A steady electric current of 3.7 A flows through a wire. How many coulombs of charge pass through the wire per minute?

60 A charge of 820 C was passed through a wire in 10 minutes. Calculate the average electric current during that interval.

61 How long does it take for 1920 C to pass a given point in a wire that carries an electric current of 1.7 A?

62 Find the resistance of a resistor through which 1920 C flow in 23 minutes if the potential difference across it is 11 V.

63 Find the electric resistance of a silver rod 46 m long, if its cross–sectional area is 1.5 mm^2 and its resistivity is 1.49×10^{-8} $\Omega \cdot$m.

64 Calculate the resistivity of a wire 40 m long, if its cross–sectional area is 2.2 mm^2 and its electric resistance is 2.17 Ω.

65 Calculate the electric resistance of an aluminum wire 2.4 mm in diameter and 34 m long if its resistivity is 2.73×10^{-8} $\Omega \cdot$m.

66 A battery with an emf (electromotive force) of 5 V delivers 143.7 mA to a 30 Ω load.
   a) What is the internal resistance of the battery?
   b) What is its terminal potential difference when joined to this load?

67 A 30 µF capacitor is charged to a potential difference of 66 V. Find the charge on the capacitor.

68 The charge on an 8 µF capacitor is 64 µC. Find the potential difference to which the capacitor was charged.

69 When a capacitor is connected across a 19 V supply the charge is 3.325 mC. Determine its capacitance.

70 A certain light bulb has a resistance of 215 Ω. If a current of 0.791 A is going through it, calculate the voltage applied.

71 A voltage of 160 V is applied to a 255 Ω resistor. Find the current through the resistor.

72 A voltage of 150 V is applied to a resistor and the current going through is found to be 0.566 A. Find the resistance of the resistor.

73 A battery with an emf (electromotive force) of 11 V has internal resistance of 1.2 Ω. Find the voltage at the terminals of the battery when 171 mA is being drawn by an external load.

74 A steady electric current of 5.3 A flows through a wire. How many coulombs of charge pass through the wire per hour?

75 A charge of 1020 C was passed through a wire in 10 minutes. Calculate the average electric current during that interval.

76 For the circuit shown below, the resistors are $R_1 = 18\ \Omega$, $R_2 = 22\ \Omega$ and $R_3 = 23\ \Omega$. The circuit applied voltage is $V_A - V_B = 47$ V. Find:

  a) The circuit's total resistance.
  b) Current flowing through resistor R_2.
  c) The voltage across R_2.

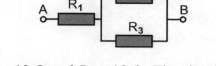

77 For the circuit shown below, the resistors are $R_1 = 29\ \Omega$, $R_2 = 13\ \Omega$ and $R_3 = 18\ \Omega$. The circuit applied voltage is $V_A - V_B = 38$ V. Find:

  a) The circuit's total resistance.
  b) Current flowing through resistor R_2.
  c) The voltage across R_2.

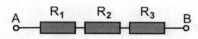

78 For the circuit shown below, the resistors are $R_1 = 14\ \Omega$, $R_2 = 18\ \Omega$ and $R_3 = 8\ \Omega$. The circuit applied voltage is $V_A - V_B = 44$ V. Find:

  a) The circuit's total resistance.
  b) Current flowing through resistor R_2.
  c) The voltage across R_2.

79 For the circuit shown below, the resistors are $R_1 = 3\ \Omega$, $R_2 = 12\ \Omega$ and $R_3 = 22\ \Omega$. The circuit applied voltage is $V_A - V_B = 19$ V. Find:

  a) The circuit's total resistance.
  b) Current flowing through resistor R_2.
  c) The voltage across R_2.

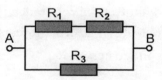

80 For the circuit shown below, the resistors are $R_1 = 11\ \Omega$, $R_2 = 15\ \Omega$, $R_3 = 22\ \Omega$ and $R_4 = 24\ \Omega$. The circuit applied voltage is $V_A - V_B = 41$ V. Find:

  a) The circuit's total resistance.
  b) Current flowing through resistor R_2.
  c) The voltage across R_2.

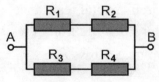

81 For the circuit shown below, the resistors are $R_1 = 11\ \Omega$, $R_2 = 24\ \Omega$, $R_3 = 25\ \Omega$, $R_4 = 21\ \Omega$ and $R_5 = 17\ \Omega$. The circuit applied voltage is $V_A - V_B = 9$ V. Find:

  a) The circuit's total resistance.
  b) Current flowing through resistor R_2.
  c) The voltage across R_2.

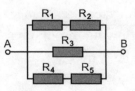

82 For the circuit shown below, the resistors are $R_1 = 6\ \Omega$ and $R_2 = 20\ \Omega$. The circuit applied voltage is $V_A - V_B = 32$ V. Find:

  a) The circuit's total resistance.
  b) Current flowing through resistor R_2.
  c) The voltage across R_2.

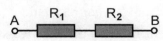

83 For the circuit shown below, the resistors are $R_1 = 15\ \Omega$ and $R_2 = 6\ \Omega$. The circuit applied voltage is $V_A - V_B = 46$ V. Find:

   a) The circuit's total resistance.
   b) Current flowing through resistor R_2.
   c) The voltage across R_2.

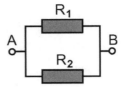

84 For the circuit shown below, the resistors are $R_1 = 9\ \Omega$, $R_2 = 12\ \Omega$ and $R_3 = 7\ \Omega$. The circuit applied voltage is $V_A - V_B = 42$ V. Find:

   a) The circuit's total resistance.
   b) Current flowing through resistor R_2.
   c) The voltage across R_2.

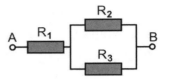

85 For the circuit shown below, the resistors are $R_1 = 24\ \Omega$, $R_2 = 29\ \Omega$ and $R_3 = 11\ \Omega$. The circuit applied voltage is $V_A - V_B = 40$ V. Find:

   a) The circuit's total resistance.
   b) Current flowing through resistor R_2.
   c) The voltage across R_2.

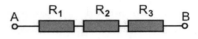

86 For the circuit shown below, the resistors are $R_1 = 29\ \Omega$, $R_2 = 28\ \Omega$ and $R_3 = 28\ \Omega$. The circuit applied voltage is $V_A - V_B = 41$ V. Find:

   a) The circuit's total resistance.
   b) Current flowing through resistor R_2.
   c) The voltage across R_2.

87 For the circuit shown below, the resistors are $R_1 = 21\ \Omega$, $R_2 = 19\ \Omega$ and $R_3 = 22\ \Omega$. The circuit applied voltage is $V_A - V_B = 22$ V. Find:

   a) The circuit's total resistance.
   b) Current flowing through resistor R_2.
   c) The voltage across R_2.

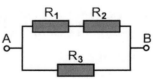

88 For the circuit shown below, the resistors are $R_1 = 4\ \Omega$, $R_2 = 21\ \Omega$, $R_3 = 30\ \Omega$ and $R_4 = 15\ \Omega$. The circuit applied voltage is $V_A - V_B = 6$ V. Find:

   a) The circuit's total resistance.
   b) Current flowing through resistor R_2.
   c) The voltage across R_2.

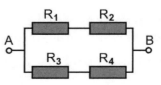

89 For the circuit shown below, the resistors are $R_1 = 28\ \Omega$, $R_2 = 28\ \Omega$, $R_3 = 24\ \Omega$, $R_4 = 16\ \Omega$ and $R_5 = 26\ \Omega$. The circuit applied voltage is $V_A - V_B = 18$ V. Find:

   a) The circuit's total resistance.
   b) Current flowing through resistor R_2.
   c) The voltage across R_2.

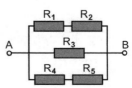

90 For the circuit shown below, the resistors are $R_1 = 12\,\Omega$ and $R_2 = 10\,\Omega$. The circuit applied voltage is $V_A - V_B = 47$ V. Find:

  a) The circuit's total resistance.
  b) Current flowing through resistor R_2.
  c) The voltage across R_2.

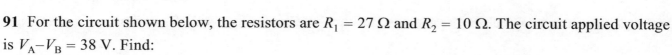

91 For the circuit shown below, the resistors are $R_1 = 27\,\Omega$ and $R_2 = 10\,\Omega$. The circuit applied voltage is $V_A - V_B = 38$ V. Find:

  a) The circuit's total resistance.
  b) Current flowing through resistor R_2.
  c) The voltage across R_2.

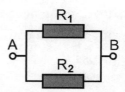

92 For the circuit shown below, the resistors are $R_1 = 17\,\Omega$, $R_2 = 24\,\Omega$ and $R_3 = 5\,\Omega$. The circuit applied voltage is $V_A - V_B = 42$ V. Find:

  a) The circuit's total resistance.
  b) Current flowing through resistor R_2.
  c) The voltage across R_2.

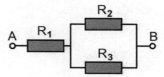

93 For the circuit shown below, the resistors are $R_1 = 26\,\Omega$, $R_2 = 22\,\Omega$ and $R_3 = 2\,\Omega$. The circuit applied voltage is $V_A - V_B = 49$ V. Find:

  a) The circuit's total resistance.
  b) Current flowing through resistor R_2.
  c) The voltage across R_2.

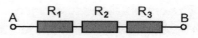

94 For the circuit shown below, the resistors are $R_1 = 10\,\Omega$, $R_2 = 6\,\Omega$ and $R_3 = 23\,\Omega$. The circuit applied voltage is $V_A - V_B = 26$ V. Find:

  a) The circuit's total resistance.
  b) Current flowing through resistor R_2.
  c) The voltage across R_2.

95 For the circuit shown below, the resistors are $R_1 = 25\,\Omega$, $R_2 = 5\,\Omega$ and $R_3 = 11\,\Omega$. The circuit applied voltage is $V_A - V_B = 16$ V. Find:

  a) The circuit's total resistance.
  b) Current flowing through resistor R_2.
  c) The voltage across R_2.

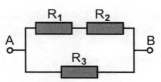

96 For the circuit shown below, the resistors are $R_1 = 15\,\Omega$, $R_2 = 20\,\Omega$, $R_3 = 22\,\Omega$ and $R_4 = 28\,\Omega$. The circuit applied voltage is $V_A - V_B = 21$ V. Find:

  a) The circuit's total resistance.
  b) Current flowing through resistor R_2.
  c) The voltage across R_2.

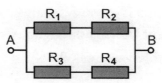

97 For the circuit shown below, the resistors are $R_1 = 27\ \Omega$, $R_2 = 26\ \Omega$, $R_3 = 8\ \Omega$, $R_4 = 27\ \Omega$ and $R_5 = 20\ \Omega$. The circuit applied voltage is $V_A-V_B = 49$ V. Find:

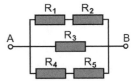

  a) The circuit's total resistance.

  b) Current flowing through resistor R_2.

  c) The voltage across R_2.

98 For the circuit shown below, the resistors are $R_1 = 20\ \Omega$ and $R_2 = 18\ \Omega$. The circuit applied voltage is $V_A-V_B = 50$ V. Find:

  a) The circuit's total resistance.

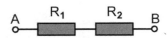

  b) Current flowing through resistor R_2.

  c) The voltage across R_2.

99 For the circuit shown below, the resistors are $R_1 = 4\ \Omega$ and $R_2 = 13\ \Omega$. The circuit applied voltage is $V_A-V_B = 18$ V. Find:

  a) The circuit's total resistance.

  b) Current flowing through resistor R_2.

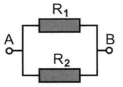

  c) The voltage across R_2.

100 For the circuit shown below, the resistors are $R_1 = 20\ \Omega$, $R_2 = 23\ \Omega$ and $R_3 = 13\ \Omega$. The circuit applied voltage is $V_A-V_B = 46$ V. Find:

  a) The circuit's total resistance.

  b) Current flowing through resistor R_2.

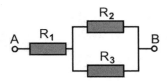

  c) The voltage across R_2.

101 For the circuit shown below, the resistors are $R_1 = 27\ \Omega$, $R_2 = 16\ \Omega$ and $R_3 = 28\ \Omega$. The circuit applied voltage is $V_A-V_B = 31$ V. Find:

  a) The circuit's total resistance.

  b) Current flowing through resistor R_2.

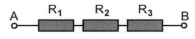

  c) The voltage across R_2.

102 For the circuit shown below, the resistors are $R_1 = 20\ \Omega$, $R_2 = 29\ \Omega$ and $R_3 = 19\ \Omega$. The circuit applied voltage is $V_A-V_B = 42$ V. Find:

  a) The circuit's total resistance.

  b) Current flowing through resistor R_2.

  c) The voltage across R_2.

103 For the circuit shown below, the resistors are $R_1 = 5\ \Omega$, $R_2 = 24\ \Omega$ and $R_3 = 25\ \Omega$. The circuit applied voltage is $V_A-V_B = 10$ V. Find:

  a) The circuit's total resistance.

  b) Current flowing through resistor R_2.

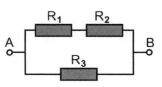

  c) The voltage across R_2.

104 For the circuit shown below, the resistors are $R_1 = 21\ \Omega$, $R_2 = 6\ \Omega$, $R_3 = 26\ \Omega$ and $R_4 = 15\ \Omega$.
The circuit applied voltage is $V_A - V_B = 38$ V. Find:
   a) The circuit's total resistance.
   b) Current flowing through resistor R_2.
   c) The voltage across R_2.

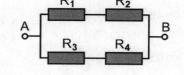

105 For the circuit shown below, the resistors are $R_1 = 15\ \Omega$, $R_2 = 25\ \Omega$, $R_3 = 29\ \Omega$, $R_4 = 11\ \Omega$ and
$R_5 = 26\ \Omega$. The circuit applied voltage is $V_A - V_B = 17$ V. Find:
   a) The circuit's total resistance.
   b) Current flowing through resistor R_2.
   c) The voltage across R_2.

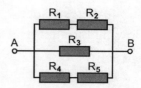

106 For the circuit shown below, the resistors are $R_1 = 10\ \Omega$ and $R_2 = 27\ \Omega$. The circuit applied
voltage is $V_A - V_B = 36$ V. Find:
   a) The circuit's total resistance.
   b) Current flowing through resistor R_2.
   c) The voltage across R_2.

107 For the circuit shown below, the resistors are $R_1 = 24\ \Omega$ and $R_2 = 5\ \Omega$. The circuit applied voltage
is $V_A - V_B = 42$ V. Find:
   a) The circuit's total resistance.
   b) Current flowing through resistor R_2.
   c) The voltage across R_2.

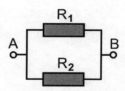

108 For the circuit shown below, the resistors are $R_1 = 29\ \Omega$, $R_2 = 21\ \Omega$ and $R_3 = 11\ \Omega$. The circuit
applied voltage is $V_A - V_B = 50$ V. Find:
   a) The circuit's total resistance.
   b) Current flowing through resistor R_2.
   c) The voltage across R_2.

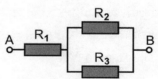

109 For the circuit shown below, the resistors are $R_1 = 20\ \Omega$, $R_2 = 29\ \Omega$ and $R_3 = 20\ \Omega$. The circuit
applied voltage is $V_A - V_B = 39$ V. Find:
   a) The circuit's total resistance.
   b) Current flowing through resistor R_2.
   c) The voltage across R_2.

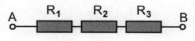

110 For the circuit shown below, the resistors are $R_1 = 2\ \Omega$, $R_2 = 20\ \Omega$ and $R_3 = 15\ \Omega$. The circuit
applied voltage is $V_A - V_B = 30$ V. Find:
   a) The circuit's total resistance.
   b) Current flowing through resistor R_2.
   c) The voltage across R_2.

111 For the circuit shown below, the resistors are $R_1 = 16\ \Omega$, $R_2 = 26\ \Omega$ and $R_3 = 5\ \Omega$. The circuit applied voltage is $V_A - V_B = 36$ V. Find:

   a) The circuit's total resistance.
   b) Current flowing through resistor R_2.
   c) The voltage across R_2.

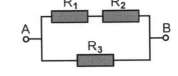

112 For the circuit shown below, the resistors are $R_1 = 4\ \Omega$, $R_2 = 19\ \Omega$, $R_3 = 19\ \Omega$ and $R_4 = 24\ \Omega$. The circuit applied voltage is $V_A - V_B = 11$ V. Find:

   a) The circuit's total resistance.
   b) Current flowing through resistor R_2.
   c) The voltage across R_2.

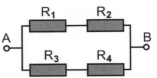

113 For the circuit shown below, the resistors are $R_1 = 11\ \Omega$, $R_2 = 28\ \Omega$, $R_3 = 21\ \Omega$, $R_4 = 26\ \Omega$ and $R_5 = 18\ \Omega$. The circuit applied voltage is $V_A - V_B = 7$ V. Find:

   a) The circuit's total resistance.
   b) Current flowing through resistor R_2.
   c) The voltage across R_2.

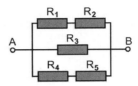

114 For the circuit shown below, the resistors are $R_1 = 7\ \Omega$ and $R_2 = 18\ \Omega$. The circuit applied voltage is $V_A - V_B = 14$ V. Find:

   a) The circuit's total resistance.
   b) Current flowing through resistor R_2.
   c) The voltage across R_2.

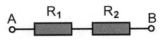

115 For the circuit shown below, the resistors are $R_1 = 24\ \Omega$ and $R_2 = 22\ \Omega$. The circuit applied voltage is $V_A - V_B = 27$ V. Find:

   a) The circuit's total resistance.
   b) Current flowing through resistor R_2.
   c) The voltage across R_2.

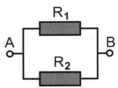

116 For the circuit shown below, the resistors are $R_1 = 5\ \Omega$, $R_2 = 8\ \Omega$ and $R_3 = 11\ \Omega$. The circuit applied voltage is $V_A - V_B = 41$ V. Find:

   a) The circuit's total resistance.
   b) Current flowing through resistor R_2.
   c) The voltage across R_2.

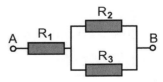

117 For the circuit shown below, the resistors are $R_1 = 28\ \Omega$, $R_2 = 14\ \Omega$ and $R_3 = 27\ \Omega$. The circuit applied voltage is $V_A - V_B = 37$ V. Find:

   a) The circuit's total resistance.
   b) Current flowing through resistor R_2.
   c) The voltage across R_2.

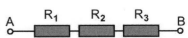

118 For the circuit shown below, the resistors are $R_1 = 14\ \Omega$, $R_2 = 3\ \Omega$ and $R_3 = 3\ \Omega$. The circuit applied voltage is $V_A - V_B = 35$ V. Find:

  a) The circuit's total resistance.

  b) Current flowing through resistor R_2.

  c) The voltage across R_2.

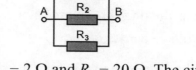

119 For the circuit shown below, the resistors are $R_1 = 19\ \Omega$, $R_2 = 2\ \Omega$ and $R_3 = 20\ \Omega$. The circuit applied voltage is $V_A - V_B = 36$ V. Find:

  a) The circuit's total resistance.

  b) Current flowing through resistor R_2.

  c) The voltage across R_2.

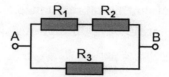

120 For the circuit shown below, the resistors are $R_1 = 11\ \Omega$, $R_2 = 21\ \Omega$, $R_3 = 14\ \Omega$ and $R_4 = 16\ \Omega$. The circuit applied voltage is $V_A - V_B = 17$ V. Find:

  a) The circuit's total resistance.

  b) Current flowing through resistor R_2.

  c) The voltage across R_2.

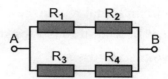

121 For the circuit shown below, the resistors are $R_1 = 30\ \Omega$, $R_2 = 9\ \Omega$, $R_3 = 2\ \Omega$, $R_4 = 3\ \Omega$ and $R_5 = 13\ \Omega$. The circuit applied voltage is $V_A - V_B = 38$ V. Find:

  a) The circuit's total resistance.

  b) Current flowing through resistor R_2.

  c) The voltage across R_2.

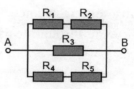

122 For the circuit shown below, the resistors are $R_1 = 29\ \Omega$ and $R_2 = 22\ \Omega$. The circuit applied voltage is $V_A - V_B = 11$ V. Find:

  a) The circuit's total resistance.

  b) Current flowing through resistor R_2.

  c) The voltage across R_2.

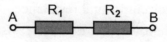

123 For the circuit shown below, the resistors are $R_1 = 13\ \Omega$ and $R_2 = 6\ \Omega$. The circuit applied voltage is $V_A - V_B = 32$ V. Find:

  a) The circuit's total resistance.

  b) Current flowing through resistor R_2.

  c) The voltage across R_2.

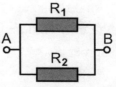

124 For the circuit shown below, the resistors are $R_1 = 28\ \Omega$, $R_2 = 5\ \Omega$ and $R_3 = 11\ \Omega$. The circuit applied voltage is $V_A - V_B = 23$ V. Find:

  a) The circuit's total resistance.

  b) Current flowing through resistor R_2.

  c) The voltage across R_2.

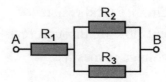

125 For the circuit shown below, the resistors are $R_1 = 6\ \Omega$, $R_2 = 15\ \Omega$ and $R_3 = 13\ \Omega$. The circuit applied voltage is $V_A - V_B = 15$ V. Find:

a) The circuit's total resistance.
b) Current flowing through resistor R_2.
c) The voltage across R_2.

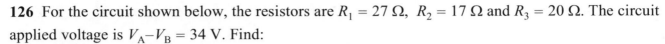

126 For the circuit shown below, the resistors are $R_1 = 27\ \Omega$, $R_2 = 17\ \Omega$ and $R_3 = 20\ \Omega$. The circuit applied voltage is $V_A - V_B = 34$ V. Find:

a) The circuit's total resistance.
b) Current flowing through resistor R_2.
c) The voltage across R_2.

127 For the circuit shown below, the resistors are $R_1 = 11\ \Omega$, $R_2 = 30\ \Omega$ and $R_3 = 12\ \Omega$. The circuit applied voltage is $V_A - V_B = 10$ V. Find:

a) The circuit's total resistance.
b) Current flowing through resistor R_2.
c) The voltage across R_2.

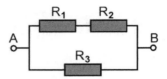

128 For the circuit shown below, the resistors are $R_1 = 26\ \Omega$, $R_2 = 29\ \Omega$, $R_3 = 28\ \Omega$ and $R_4 = 16\ \Omega$. The circuit applied voltage is $V_A - V_B = 46$ V. Find:

a) The circuit's total resistance.
b) Current flowing through resistor R_2.
c) The voltage across R_2.

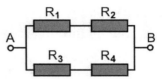

129 For the circuit shown below, the resistors are $R_1 = 8\ \Omega$, $R_2 = 3\ \Omega$, $R_3 = 4\ \Omega$, $R_4 = 5\ \Omega$ and $R_5 = 16\ \Omega$. The circuit applied voltage is $V_A - V_B = 7$ V. Find:

a) The circuit's total resistance.
b) Current flowing through resistor R_2.
c) The voltage across R_2.

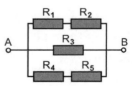

130 For the circuit shown below, the resistors are $R_1 = 6\ \Omega$ and $R_2 = 14\ \Omega$. The circuit applied voltage is $V_A - V_B = 29$ V. Find:

a) The circuit's total resistance.
b) Current flowing through resistor R_2.
c) The voltage across R_2.

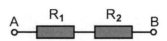

131 For the circuit shown below, the resistors are $R_1 = 15\ \Omega$ and $R_2 = 22\ \Omega$. The circuit applied voltage is $V_A - V_B = 28$ V. Find:

a) The circuit's total resistance.
b) Current flowing through resistor R_2.
c) The voltage across R_2.

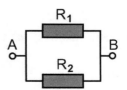

132 For the circuit shown below, the resistors are $R_1 = 28\ \Omega$, $R_2 = 17\ \Omega$ and $R_3 = 23\ \Omega$. The circuit applied voltage is $V_A - V_B = 42$ V. Find:

   a) The circuit's total resistance.

   b) Current flowing through resistor R_2.

   c) The voltage across R_2.

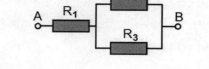

133 For the circuit shown below, the resistors are $R_1 = 16\ \Omega$, $R_2 = 28\ \Omega$ and $R_3 = 23\ \Omega$. The circuit applied voltage is $V_A - V_B = 24$ V. Find:

   a) The circuit's total resistance.

   b) Current flowing through resistor R_2.

   c) The voltage across R_2.

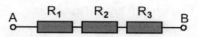

134 For the circuit shown below, the resistors are $R_1 = 6\ \Omega$, $R_2 = 25\ \Omega$ and $R_3 = 19\ \Omega$. The circuit applied voltage is $V_A - V_B = 21$ V. Find:

   a) The circuit's total resistance.

   b) Current flowing through resistor R_2.

   c) The voltage across R_2.

135 For the circuit shown below, the resistors are $R_1 = 16\ \Omega$, $R_2 = 5\ \Omega$ and $R_3 = 30\ \Omega$. The circuit applied voltage is $V_A - V_B = 21$ V. Find:

   a) The circuit's total resistance.

   b) Current flowing through resistor R_2.

   c) The voltage across R_2.

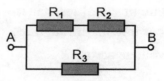

136 For the circuit shown below, the resistors are $R_1 = 29\ \Omega$, $R_2 = 28\ \Omega$, $R_3 = 7\ \Omega$ and $R_4 = 4\ \Omega$. The circuit applied voltage is $V_A - V_B = 34$ V. Find:

   a) The circuit's total resistance.

   b) Current flowing through resistor R_2.

   c) The voltage across R_2.

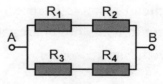

137 For the circuit shown below, the resistors are $R_1 = 24\ \Omega$, $R_2 = 2\ \Omega$, $R_3 = 23\ \Omega$, $R_4 = 21\ \Omega$ and $R_5 = 5\ \Omega$. The circuit applied voltage is $V_A - V_B = 50$ V. Find:

   a) The circuit's total resistance.

   b) Current flowing through resistor R_2.

   c) The voltage across R_2.

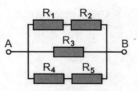

138 For the circuit shown below, the resistors are $R_1 = 15\ \Omega$ and $R_2 = 23\ \Omega$. The circuit applied voltage is $V_A - V_B = 25$ V. Find:

   a) The circuit's total resistance.

   b) Current flowing through resistor R_2.

   c) The voltage across R_2.

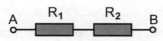

139 For the circuit shown below, the resistors are $R_1 = 9 \ \Omega$ and $R_2 = 19 \ \Omega$. The circuit applied voltage is $V_A - V_B = 29$ V. Find:

  a) The circuit's total resistance.
  b) Current flowing through resistor R_2.
  c) The voltage across R_2.

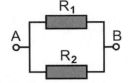

140 For the circuit shown below, the resistors are $R_1 = 23 \ \Omega$, $R_2 = 24 \ \Omega$ and $R_3 = 12 \ \Omega$. The circuit applied voltage is $V_A - V_B = 34$ V. Find:

  a) The circuit's total resistance.
  b) Current flowing through resistor R_2.
  c) The voltage across R_2.

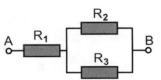

141 For the circuit shown below, the resistors are $R_1 = 11 \ \Omega$, $R_2 = 18 \ \Omega$ and $R_3 = 25 \ \Omega$. The circuit applied voltage is $V_A - V_B = 9$ V. Find:

  a) The circuit's total resistance.
  b) Current flowing through resistor R_2.
  c) The voltage across R_2.

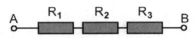

142 For the circuit shown below, the resistors are $R_1 = 19 \ \Omega$, $R_2 = 21 \ \Omega$ and $R_3 = 25 \ \Omega$. The circuit applied voltage is $V_A - V_B = 32$ V. Find:

  a) The circuit's total resistance.
  b) Current flowing through resistor R_2.
  c) The voltage across R_2.

143 For the circuit shown below, the resistors are $R_1 = 29 \ \Omega$, $R_2 = 12 \ \Omega$ and $R_3 = 21 \ \Omega$. The circuit applied voltage is $V_A - V_B = 23$ V. Find:

  a) The circuit's total resistance.
  b) Current flowing through resistor R_2.
  c) The voltage across R_2.

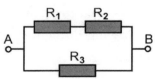

144 For the circuit shown below, the resistors are $R_1 = 15 \ \Omega$, $R_2 = 18 \ \Omega$, $R_3 = 11 \ \Omega$ and $R_4 = 23 \ \Omega$. The circuit applied voltage is $V_A - V_B = 29$ V. Find:

  a) The circuit's total resistance.
  b) Current flowing through resistor R_2.
  c) The voltage across R_2.

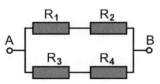

145 For the circuit shown below, the resistors are $R_1 = 25 \ \Omega$, $R_2 = 25 \ \Omega$, $R_3 = 21 \ \Omega$, $R_4 = 23 \ \Omega$ and $R_5 = 2 \ \Omega$. The circuit applied voltage is $V_A - V_B = 31$ V. Find:

  a) The circuit's total resistance.
  b) Current flowing through resistor R_2.
  c) The voltage across R_2.

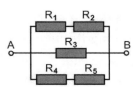

146 For the circuit shown below, the resistors are $R_1 = 23\ \Omega$ and $R_2 = 8\ \Omega$. The circuit applied voltage is $V_A - V_B = 23$ V. Find:

   a) The circuit's total resistance.

   b) Current flowing through resistor R_2.

   c) The voltage across R_2.

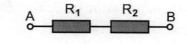

147 For the circuit shown below, the resistors are $R_1 = 9\ \Omega$ and $R_2 = 25\ \Omega$. The circuit applied voltage is $V_A - V_B = 22$ V. Find:

   a) The circuit's total resistance.

   b) Current flowing through resistor R_2.

   c) The voltage across R_2.

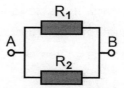

148 For the circuit shown below, the resistors are $R_1 = 26\ \Omega$, $R_2 = 29\ \Omega$ and $R_3 = 27\ \Omega$. The circuit applied voltage is $V_A - V_B = 34$ V. Find:

   a) The circuit's total resistance.

   b) Current flowing through resistor R_2.

   c) The voltage across R_2.

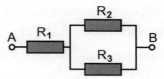

149 For the circuit shown below, the resistors are $R_1 = 7\ \Omega$, $R_2 = 15\ \Omega$ and $R_3 = 11\ \Omega$. The circuit applied voltage is $V_A - V_B = 48$ V. Find:

   a) The circuit's total resistance.

   b) Current flowing through resistor R_2.

   c) The voltage across R_2.

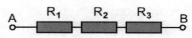

150 For the circuit shown below, the resistors are $R_1 = 18\ \Omega$, $R_2 = 16\ \Omega$ and $R_3 = 19\ \Omega$. The circuit applied voltage is $V_A - V_B = 32$ V. Find:

   a) The circuit's total resistance.

   b) Current flowing through resistor R_2.

   c) The voltage across R_2.

151 For the circuit shown below the capacitances are $C_1 = 4\ \mu F$, $C_2 = 5\ \mu F$ and $C_3 = 23\ \mu F$. The circuit applied voltage is $V_A - V_B = 30$ V. Find:

   a) Circuit's equivalent capacitance.

   b) Charge stored on capacitor C_2.

   c) Potential difference across C_2.

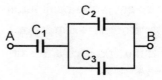

152 For the circuit shown below the capacitances are $C_1 = 21\ \mu F$, $C_2 = 9\ \mu F$ and $C_3 = 21\ \mu F$. The circuit applied voltage is $V_A - V_B = 43$ V. Find:

   a) Circuit's equivalent capacitance.

   b) Charge stored on capacitor C_2.

   c) Potential difference across C_2.

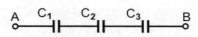

153 For the circuit shown below the capacitances are $C_1 = 13$ μF, $C_2 = 28$ μF and $C_3 = 15$ μF. The circuit applied voltage is $V_A - V_B = 24$ V. Find:

  a) Circuit's equivalent capacitance.
  b) Charge stored on capacitor C_2.
  c) Potential difference across C_2.

154 For the circuit shown below the capacitances are $C_1 = 24$ μF, $C_2 = 30$ μF and $C_3 = 23$ μF. The circuit applied voltage is $V_A - V_B = 8$ V. Find:

  a) Circuit's equivalent capacitance.
  b) Charge stored on capacitor C_2.
  c) Potential difference across C_2.

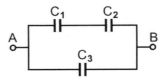

155 For the circuit shown below the capacitances are $C_1 = 10$ μF, $C_2 = 18$ μF, $C_3 = 26$ μF and $C_4 = 14$ μF. The circuit applied voltage is $V_A - V_B = 6$ V. Find:

  a) Circuit's equivalent capacitance.
  b) Charge stored on capacitor C_2.
  c) Potential difference across C_2.

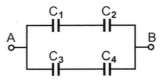

156 For the circuit shown below the capacitances are $C_1 = 2$ μF, $C_2 = 28$ μF, $C_3 = 12$ μF, $C_4 = 16$ μF and $C_5 = 28$ μF. The circuit applied voltage is $V_A - V_B = 12$ V. Find:

  a) Circuit's equivalent capacitance.
  b) Charge stored on capacitor C_2.
  c) Potential difference across C_2.

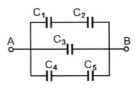

157 For the circuit shown below the capacitances are $C_1 = 24$ μF and $C_2 = 25$ μF. The circuit applied voltage is $V_A - V_B = 9$ V. Find:

  a) Circuit's equivalent capacitance.
  b) Charge stored on capacitor C_2.
  c) Potential difference across C_2.

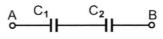

158 For the circuit shown below the capacitances are $C_1 = 13$ μF and $C_2 = 12$ μF. The circuit applied voltage is $V_A - V_B = 13$ V. Find:

  a) Circuit's equivalent capacitance.
  b) Charge stored on capacitor C_2.
  c) Potential difference across C_2.

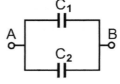

159 For the circuit shown below the capacitances are $C_1 = 22$ μF, $C_2 = 9$ μF and $C_3 = 25$ μF. The circuit applied voltage is $V_A - V_B = 20$ V. Find:

  a) Circuit's equivalent capacitance.
  b) Charge stored on capacitor C_2.
  c) Potential difference across C_2.

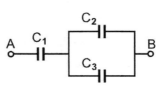

160 For the circuit shown below the capacitances are $C_1 = 18\ \mu F$, $C_2 = 26\ \mu F$ and $C_3 = 27\ \mu F$. The circuit applied voltage is $V_A - V_B = 8$ V. Find:

a) Circuit's equivalent capacitance.
b) Charge stored on capacitor C_2.
c) Potential difference across C_2.

161 For the circuit shown below the capacitances are $C_1 = 21\ \mu F$, $C_2 = 21\ \mu F$ and $C_3 = 6\ \mu F$. The circuit applied voltage is $V_A - V_B = 32$ V. Find:

a) Circuit's equivalent capacitance.
b) Charge stored on capacitor C_2.
c) Potential difference across C_2.

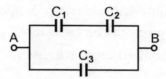

162 For the circuit shown below the capacitances are $C_1 = 11\ \mu F$, $C_2 = 7\ \mu F$ and $C_3 = 17\ \mu F$. The circuit applied voltage is $V_A - V_B = 36$ V. Find:

a) Circuit's equivalent capacitance.
b) Charge stored on capacitor C_2.
c) Potential difference across C_2.

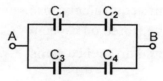

163 For the circuit shown below the capacitances are $C_1 = 4\ \mu F$, $C_2 = 9\ \mu F$, $C_3 = 24\ \mu F$ and $C_4 = 12\ \mu F$. The circuit applied voltage is $V_A - V_B = 10$ V. Find:

a) Circuit's equivalent capacitance.
b) Charge stored on capacitor C_2.
c) Potential difference across C_2.

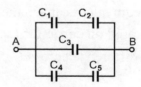

164 For the circuit shown below the capacitances are $C_1 = 9\ \mu F$, $C_2 = 2\ \mu F$, $C_3 = 7\ \mu F$, $C_4 = 9\ \mu F$ and $C_5 = 11\ \mu F$. The circuit applied voltage is $V_A - V_B = 18$ V. Find:

a) Circuit's equivalent capacitance.
b) Charge stored on capacitor C_2.
c) Potential difference across C_2.

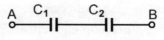

165 For the circuit shown below the capacitances are $C_1 = 7\ \mu F$ and $C_2 = 3\ \mu F$. The circuit applied voltage is $V_A - V_B = 18$ V. Find:

a) Circuit's equivalent capacitance.
b) Charge stored on capacitor C_2.
c) Potential difference across C_2.

166 For the circuit shown below the capacitances are $C_1 = 25\ \mu F$ and $C_2 = 29\ \mu F$. The circuit applied voltage is $V_A - V_B = 44$ V. Find:

a) Circuit's equivalent capacitance.
b) Charge stored on capacitor C_2.
c) Potential difference across C_2.

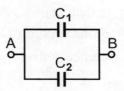

167 For the circuit shown below the capacitances are $C_1 = 15$ µF, $C_2 = 26$ µF and $C_3 = 14$ µF. The circuit applied voltage is $V_A - V_B = 41$ V. Find:

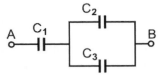

a) Circuit's equivalent capacitance.
b) Charge stored on capacitor C_2.
c) Potential difference across C_2.

168 For the circuit shown below the capacitances are $C_1 = 25$ µF, $C_2 = 6$ µF and $C_3 = 25$ µF. The circuit applied voltage is $V_A - V_B = 41$ V. Find:

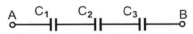

a) Circuit's equivalent capacitance.
b) Charge stored on capacitor C_2.
c) Potential difference across C_2.

169 For the circuit shown below the capacitances are $C_1 = 21$ µF, $C_2 = 6$ µF and $C_3 = 11$ µF. The circuit applied voltage is $V_A - V_B = 21$ V. Find:

a) Circuit's equivalent capacitance.
b) Charge stored on capacitor C_2.
c) Potential difference across C_2.

170 For the circuit shown below the capacitances are $C_1 = 26$ µF, $C_2 = 12$ µF and $C_3 = 3$ µF. The circuit applied voltage is $V_A - V_B = 37$ V. Find:

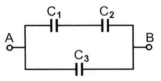

a) Circuit's equivalent capacitance.
b) Charge stored on capacitor C_2.
c) Potential difference across C_2.

171 For the circuit shown below the capacitances are $C_1 = 16$ µF, $C_2 = 9$ µF, $C_3 = 27$ µF and $C_4 = 14$ µF. The circuit applied voltage is $V_A - V_B = 34$ V. Find:

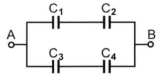

a) Circuit's equivalent capacitance.
b) Charge stored on capacitor C_2.
c) Potential difference across C_2.

172 For the circuit shown below the capacitances are $C_1 = 12$ µF, $C_2 = 5$ µF, $C_3 = 12$ µF, $C_4 = 20$ µF and $C_5 = 12$ µF. The circuit applied voltage is $V_A - V_B = 34$ V. Find:

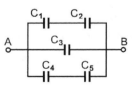

a) Circuit's equivalent capacitance.
b) Charge stored on capacitor C_2.
c) Potential difference across C_2.

173 For the circuit shown below the capacitances are $C_1 = 6$ µF and $C_2 = 13$ µF. The circuit applied voltage is $V_A - V_B = 24$ V. Find:

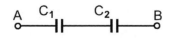

a) Circuit's equivalent capacitance.
b) Charge stored on capacitor C_2.
c) Potential difference across C_2.

174 For the circuit shown below the capacitances are $C_1 = 23\ \mu F$ and $C_2 = 6\ \mu F$. The circuit applied voltage is $V_A - V_B = 44$ V. Find:

   a) Circuit's equivalent capacitance.

   b) Charge stored on capacitor C_2.

   c) Potential difference across C_2.

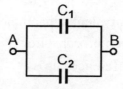

175 For the circuit shown below the capacitances are $C_1 = 3\ \mu F$, $C_2 = 9\ \mu F$ and $C_3 = 16\ \mu F$. The circuit applied voltage is $V_A - V_B = 21$ V. Find:

   a) Circuit's equivalent capacitance.

   b) Charge stored on capacitor C_2.

   c) Potential difference across C_2.

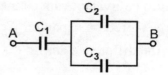

176 For the circuit shown below the capacitances are $C_1 = 25\ \mu F$, $C_2 = 5\ \mu F$ and $C_3 = 20\ \mu F$. The circuit applied voltage is $V_A - V_B = 24$ V. Find:

   a) Circuit's equivalent capacitance.

   b) Charge stored on capacitor C_2.

   c) Potential difference across C_2.

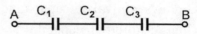

177 For the circuit shown below the capacitances are $C_1 = 10\ \mu F$, $C_2 = 11\ \mu F$ and $C_3 = 10\ \mu F$. The circuit applied voltage is $V_A - V_B = 30$ V. Find:

   a) Circuit's equivalent capacitance.

   b) Charge stored on capacitor C_2.

   c) Potential difference across C_2.

178 For the circuit shown below the capacitances are $C_1 = 15\ \mu F$, $C_2 = 27\ \mu F$ and $C_3 = 15\ \mu F$. The circuit applied voltage is $V_A - V_B = 38$ V. Find:

   a) Circuit's equivalent capacitance.

   b) Charge stored on capacitor C_2.

   c) Potential difference across C_2.

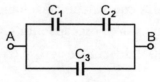

179 For the circuit shown below the capacitances are $C_1 = 6\ \mu F$, $C_2 = 7\ \mu F$, $C_3 = 24\ \mu F$ and $C_4 = 25\ \mu F$. The circuit applied voltage is $V_A - V_B = 27$ V. Find:

   a) Circuit's equivalent capacitance.

   b) Charge stored on capacitor C_2.

   c) Potential difference across C_2.

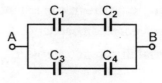

180 For the circuit shown below the capacitances are $C_1 = 12\ \mu F$, $C_2 = 6\ \mu F$, $C_3 = 13\ \mu F$, $C_4 = 5\ \mu F$ and $C_5 = 28\ \mu F$. The circuit applied voltage is $V_A - V_B = 31$ V. Find:

   a) Circuit's equivalent capacitance.

   b) Charge stored on capacitor C_2.

   c) Potential difference across C_2.

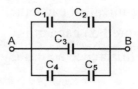

181 For the circuit shown below the capacitances are $C_1 = 13\ \mu F$ and $C_2 = 4\ \mu F$. The circuit applied voltage is $V_A - V_B = 37$ V. Find:

   a) Circuit's equivalent capacitance.
   b) Charge stored on capacitor C_2.
   c) Potential difference across C_2.

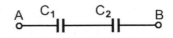

182 For the circuit shown below the capacitances are $C_1 = 21\ \mu F$ and $C_2 = 19\ \mu F$. The circuit applied voltage is $V_A - V_B = 7$ V. Find:

   a) Circuit's equivalent capacitance.
   b) Charge stored on capacitor C_2.
   c) Potential difference across C_2.

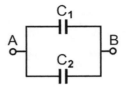

183 For the circuit shown below the capacitances are $C_1 = 5\ \mu F$, $C_2 = 20\ \mu F$ and $C_3 = 5\ \mu F$. The circuit applied voltage is $V_A - V_B = 36$ V. Find:

   a) Circuit's equivalent capacitance.
   b) Charge stored on capacitor C_2.
   c) Potential difference across C_2.

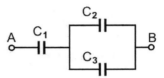

184 For the circuit shown below the capacitances are $C_1 = 19\ \mu F$, $C_2 = 9\ \mu F$ and $C_3 = 8\ \mu F$. The circuit applied voltage is $V_A - V_B = 34$ V. Find:

   a) Circuit's equivalent capacitance.
   b) Charge stored on capacitor C_2.
   c) Potential difference across C_2.

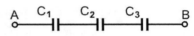

185 For the circuit shown below the capacitances are $C_1 = 8\ \mu F$, $C_2 = 13\ \mu F$ and $C_3 = 6\ \mu F$. The circuit applied voltage is $V_A - V_B = 46$ V. Find:

   a) Circuit's equivalent capacitance.
   b) Charge stored on capacitor C_2.
   c) Potential difference across C_2.

186 For the circuit shown below the capacitances are $C_1 = 28\ \mu F$, $C_2 = 4\ \mu F$ and $C_3 = 28\ \mu F$. The circuit applied voltage is $V_A - V_B = 28$ V. Find:

   a) Circuit's equivalent capacitance.
   b) Charge stored on capacitor C_2.
   c) Potential difference across C_2.

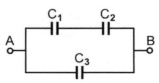

187 For the circuit shown below the capacitances are $C_1 = 11\ \mu F$, $C_2 = 5\ \mu F$, $C_3 = 8\ \mu F$ and $C_4 = 7\ \mu F$. The circuit applied voltage is $V_A - V_B = 12$ V. Find:

   a) Circuit's equivalent capacitance.
   b) Charge stored on capacitor C_2.
   c) Potential difference across C_2.

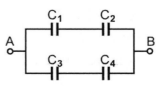

188 For the circuit shown below the capacitances are $C_1 = 9$ µF, $C_2 = 28$ µF, $C_3 = 25$ µF, $C_4 = 12$ µF and $C_5 = 10$ µF. The circuit applied voltage is $V_A - V_B = 38$ V. Find:

   a) Circuit's equivalent capacitance.
   b) Charge stored on capacitor C_2.
   c) Potential difference across C_2.

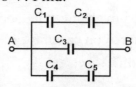

189 For the circuit shown below the capacitances are $C_1 = 26$ µF and $C_2 = 16$ µF. The circuit applied voltage is $V_A - V_B = 36$ V. Find:

   a) Circuit's equivalent capacitance.
   b) Charge stored on capacitor C_2.
   c) Potential difference across C_2.

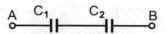

190 For the circuit shown below the capacitances are $C_1 = 13$ µF and $C_2 = 16$ µF. The circuit applied voltage is $V_A - V_B = 27$ V. Find:

   a) Circuit's equivalent capacitance.
   b) Charge stored on capacitor C_2.
   c) Potential difference across C_2.

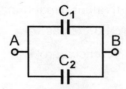

191 For the circuit shown below the capacitances are $C_1 = 3$ µF, $C_2 = 3$ µF and $C_3 = 8$ µF. The circuit applied voltage is $V_A - V_B = 31$ V. Find:

   a) Circuit's equivalent capacitance.
   b) Charge stored on capacitor C_2.
   c) Potential difference across C_2.

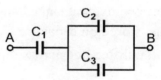

192 For the circuit shown below the capacitances are $C_1 = 12$ µF, $C_2 = 28$ µF and $C_3 = 6$ µF. The circuit applied voltage is $V_A - V_B = 38$ V. Find:

   a) Circuit's equivalent capacitance.
   b) Charge stored on capacitor C_2.
   c) Potential difference across C_2.

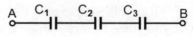

193 For the circuit shown below the capacitances are $C_1 = 24$ µF, $C_2 = 12$ µF and $C_3 = 27$ µF. The circuit applied voltage is $V_A - V_B = 14$ V. Find:

   a) Circuit's equivalent capacitance.
   b) Charge stored on capacitor C_2.
   c) Potential difference across C_2.

194 For the circuit shown below the capacitances are $C_1 = 11$ µF, $C_2 = 11$ µF and $C_3 = 15$ µF. The circuit applied voltage is $V_A - V_B = 50$ V. Find:

   a) Circuit's equivalent capacitance.
   b) Charge stored on capacitor C_2.
   c) Potential difference across C_2.

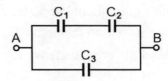

195 For the circuit shown below the capacitances are $C_1 = 29$ μF, $C_2 = 26$ μF, $C_3 = 26$ μF and $C_4 = 7$ μF. The circuit applied voltage is $V_A - V_B = 9$ V. Find:

   a) Circuit's equivalent capacitance.
   b) Charge stored on capacitor C_2.
   c) Potential difference across C_2.

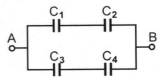

196 For the circuit shown below the capacitances are $C_1 = 4$ μF, $C_2 = 15$ μF, $C_3 = 26$ μF, $C_4 = 22$ μF and $C_5 = 23$ μF. The circuit applied voltage is $V_A - V_B = 40$ V. Find:

   a) Circuit's equivalent capacitance.
   b) Charge stored on capacitor C_2.
   c) Potential difference across C_2.

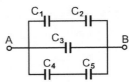

197 For the circuit shown below the capacitances are $C_1 = 3$ μF and $C_2 = 9$ μF. The circuit applied voltage is $V_A - V_B = 24$ V. Find:

   a) Circuit's equivalent capacitance.
   b) Charge stored on capacitor C_2.
   c) Potential difference across C_2.

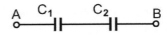

198 For the circuit shown below the capacitances are $C_1 = 16$ μF and $C_2 = 4$ μF. The circuit applied voltage is $V_A - V_B = 38$ V. Find:

   a) Circuit's equivalent capacitance.
   b) Charge stored on capacitor C_2.
   c) Potential difference across C_2.

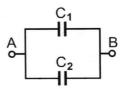

199 For the circuit shown below the capacitances are $C_1 = 3$ μF, $C_2 = 14$ μF and $C_3 = 22$ μF. The circuit applied voltage is $V_A - V_B = 20$ V. Find:

   a) Circuit's equivalent capacitance.
   b) Charge stored on capacitor C_2.
   c) Potential difference across C_2.

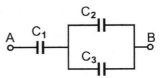

200 For the circuit shown below the capacitances are $C_1 = 22$ μF, $C_2 = 12$ μF and $C_3 = 18$ μF. The circuit applied voltage is $V_A - V_B = 17$ V. Find:

   a) Circuit's equivalent capacitance.
   b) Charge stored on capacitor C_2.
   c) Potential difference across C_2.

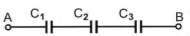

201 For the circuit shown below the capacitances are $C_1 = 12$ μF, $C_2 = 26$ μF and $C_3 = 2$ μF. The circuit applied voltage is $V_A - V_B = 37$ V. Find:

   a) Circuit's equivalent capacitance.
   b) Charge stored on capacitor C_2.
   c) Potential difference across C_2.

202 For the circuit shown below the capacitances are $C_1 = 24$ μF, $C_2 = 21$ μF and $C_3 = 8$ μF. The circuit applied voltage is $V_A - V_B = 5$ V. Find:

a) Circuit's equivalent capacitance.
b) Charge stored on capacitor C_2.
c) Potential difference across C_2.

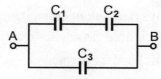

203 For the circuit shown below the capacitances are $C_1 = 12$ μF, $C_2 = 3$ μF, $C_3 = 9$ μF and $C_4 = 4$ μF. The circuit applied voltage is $V_A - V_B = 33$ V. Find:

a) Circuit's equivalent capacitance.
b) Charge stored on capacitor C_2.
c) Potential difference across C_2.

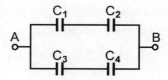

204 For the circuit shown below the capacitances are $C_1 = 9$ μF, $C_2 = 10$ μF, $C_3 = 7$ μF, $C_4 = 3$ μF and $C_5 = 16$ μF. The circuit applied voltage is $V_A - V_B = 26$ V. Find:

a) Circuit's equivalent capacitance.
b) Charge stored on capacitor C_2.
c) Potential difference across C_2.

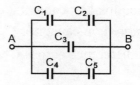

205 For the circuit shown below the capacitances are $C_1 = 27$ μF and $C_2 = 9$ μF. The circuit applied voltage is $V_A - V_B = 16$ V. Find:

a) Circuit's equivalent capacitance.
b) Charge stored on capacitor C_2.
c) Potential difference across C_2.

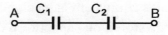

206 For the circuit shown below the capacitances are $C_1 = 12$ μF and $C_2 = 15$ μF. The circuit applied voltage is $V_A - V_B = 22$ V. Find:

a) Circuit's equivalent capacitance.
b) Charge stored on capacitor C_2.
c) Potential difference across C_2.

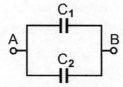

207 For the circuit shown below the capacitances are $C_1 = 29$ μF, $C_2 = 8$ μF and $C_3 = 7$ μF. The circuit applied voltage is $V_A - V_B = 42$ V. Find:

a) Circuit's equivalent capacitance.
b) Charge stored on capacitor C_2.
c) Potential difference across C_2.

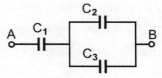

208 For the circuit shown below the capacitances are $C_1 = 8$ μF, $C_2 = 11$ μF and $C_3 = 7$ μF. The circuit applied voltage is $V_A - V_B = 19$ V. Find:

a) Circuit's equivalent capacitance.
b) Charge stored on capacitor C_2.
c) Potential difference across C_2.

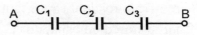

209 For the circuit shown below the capacitances are $C_1 = 2$ µF, $C_2 = 7$ µF and $C_3 = 11$ µF. The circuit applied voltage is $V_A - V_B = 23$ V. Find:

a) Circuit's equivalent capacitance.
b) Charge stored on capacitor C_2.
c) Potential difference across C_2.

210 For the circuit shown below the capacitances are $C_1 = 16$ µF, $C_2 = 13$ µF and $C_3 = 30$ µF. The circuit applied voltage is $V_A - V_B = 29$ V. Find:

a) Circuit's equivalent capacitance.
b) Charge stored on capacitor C_2.
c) Potential difference across C_2.

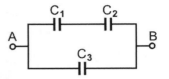

211 For the circuit shown below the capacitances are $C_1 = 22$ µF, $C_2 = 25$ µF, $C_3 = 22$ µF and $C_4 = 5$ µF. The circuit applied voltage is $V_A - V_B = 6$ V. Find:

a) Circuit's equivalent capacitance.
b) Charge stored on capacitor C_2.
c) Potential difference across C_2.

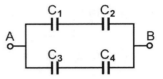

212 For the circuit shown below the capacitances are $C_1 = 3$ µF, $C_2 = 22$ µF, $C_3 = 20$ µF, $C_4 = 11$ µF and $C_5 = 7$ µF. The circuit applied voltage is $V_A - V_B = 17$ V. Find:

a) Circuit's equivalent capacitance.
b) Charge stored on capacitor C_2.
c) Potential difference across C_2.

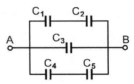

213 For the circuit shown below the capacitances are $C_1 = 22$ µF and $C_2 = 9$ µF. The circuit applied voltage is $V_A - V_B = 41$ V. Find:

a) Circuit's equivalent capacitance.
b) Charge stored on capacitor C_2.
c) Potential difference across C_2.

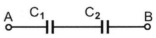

214 For the circuit shown below the capacitances are $C_1 = 11$ µF and $C_2 = 13$ µF. The circuit applied voltage is $V_A - V_B = 8$ V. Find:

a) Circuit's equivalent capacitance.
b) Charge stored on capacitor C_2.
c) Potential difference across C_2.

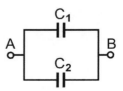

215 For the circuit shown below the capacitances are $C_1 = 19$ µF, $C_2 = 27$ µF and $C_3 = 19$ µF. The circuit applied voltage is $V_A - V_B = 22$ V. Find:

a) Circuit's equivalent capacitance.
b) Charge stored on capacitor C_2.
c) Potential difference across C_2.

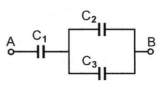

216 For the circuit shown below the capacitances are $C_1 = 3$ μF, $C_2 = 28$ μF and $C_3 = 27$ μF. The circuit applied voltage is $V_A - V_B = 6$ V. Find:

 a) Circuit's equivalent capacitance.
 b) Charge stored on capacitor C_2.
 c) Potential difference across C_2.

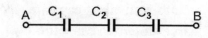

217 For the circuit shown below the capacitances are $C_1 = 17$ μF, $C_2 = 17$ μF and $C_3 = 2$ μF. The circuit applied voltage is $V_A - V_B = 49$ V. Find:

 a) Circuit's equivalent capacitance.
 b) Charge stored on capacitor C_2.
 c) Potential difference across C_2.

218 For the circuit shown below the capacitances are $C_1 = 6$ μF, $C_2 = 21$ μF and $C_3 = 28$ μF. The circuit applied voltage is $V_A - V_B = 17$ V. Find:

 a) Circuit's equivalent capacitance.
 b) Charge stored on capacitor C_2.
 c) Potential difference across C_2.

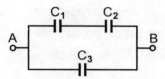

219 For the circuit shown below the capacitances are $C_1 = 26$ μF, $C_2 = 12$ μF, $C_3 = 22$ μF and $C_4 = 21$ μF. The circuit applied voltage is $V_A - V_B = 43$ V. Find:

 a) Circuit's equivalent capacitance.
 b) Charge stored on capacitor C_2.
 c) Potential difference across C_2.

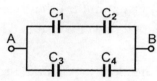

220 For the circuit shown below the capacitances are $C_1 = 19$ μF, $C_2 = 15$ μF, $C_3 = 6$ μF, $C_4 = 12$ μF and $C_5 = 11$ μF. The circuit applied voltage is $V_A - V_B = 33$ V. Find:

 a) Circuit's equivalent capacitance.
 b) Charge stored on capacitor C_2.
 c) Potential difference across C_2.

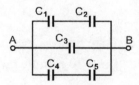

221 For the circuit shown below the capacitances are $C_1 = 15$ μF and $C_2 = 28$ μF. The circuit applied voltage is $V_A - V_B = 45$ V. Find:

 a) Circuit's equivalent capacitance.
 b) Charge stored on capacitor C_2.
 c) Potential difference across C_2.

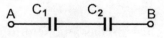

222 For the circuit shown below the capacitances are $C_1 = 27$ μF and $C_2 = 12$ μF. The circuit applied voltage is $V_A - V_B = 21$ V. Find:

 a) Circuit's equivalent capacitance.
 b) Charge stored on capacitor C_2.
 c) Potential difference across C_2.

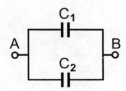

223 For the circuit shown below the capacitances are $C_1 = 14$ μF, $C_2 = 11$ μF and $C_3 = 14$ μF. The circuit applied voltage is $V_A - V_B = 11$ V. Find:

a) Circuit's equivalent capacitance.

b) Charge stored on capacitor C_2.

c) Potential difference across C_2.

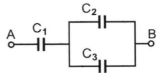

224 For the circuit shown below the capacitances are $C_1 = 21$ μF, $C_2 = 21$ μF and $C_3 = 17$ μF. The circuit applied voltage is $V_A - V_B = 49$ V. Find:

a) Circuit's equivalent capacitance.

b) Charge stored on capacitor C_2.

c) Potential difference across C_2.

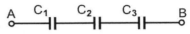

225 For the circuit shown below the capacitances are $C_1 = 13$ μF, $C_2 = 23$ μF and $C_3 = 23$ μF. The circuit applied voltage is $V_A - V_B = 23$ V. Find:

a) Circuit's equivalent capacitance.

b) Charge stored on capacitor C_2.

c) Potential difference across C_2.

226 Two resistors, 27 Ω and 40 Ω, are connected in parallel. If the current through the 27 Ω resistor is 1.81 A, find:

a) The current in the other resistor.

b) The total power consumed by the two resistors.

227 Find the electric potential drop across a 32 W resistor that draws a current of 4 A.

228 A 66 W electric lamp draws a current of 2.9 A. Find the voltage applied and the resistance of the lamp.

229 A 6 kW electric heater operates at 550 V. Find the current, resistance, and energy (in joules and kW·h) generated in 120 minutes.

230 The operating potential difference of a light bulb is 205 V and its power rating is 90 W. Calculate the current in the bulb and its resistance.

231 A light bulb has an operating potential difference of 35 V and draws a current of 5 A. Calculate its power rating and its resistance.

232 How much does it cost to operate a 98 W light bulb for 14 days if electrical energy costs $0.129 per kW·h?

233 An 81 W lamp is operated for 13 hours a day. How much energy does it take to operate the lamp for 7 days? Express your answer in joules and kilowatt–hours.

234 A resistor is rated at 365 W and its electrical resistance is 1290 Ω.
   a) What is the maximum voltage that can be applied across the resistor?
   b) What is the current at this voltage?

235 Three resistors, 10 Ω, 23 Ω and 34 Ω, are connected in series across a battery that provides 38 V.
   a) Calculate the voltage across each resistor.
   b) Calculate the power dissipated in each resistor.

236 Two resistors, 30 Ω and 39 Ω, are connected in parallel. If the current through the 30 Ω resistor is 2.77 A, find:
   a) The current in the other resistor.
   b) The total power consumed by the two resistors.

237 Find the electric potential drop across an 86 W resistor that draws a current of 4.7 A.

238 An 80 W electric lamp draws a current of 840 mA. Find the voltage applied and the resistance of the lamp.

239 A 30 kW electric heater operates at 575 V. Find the current, resistance, and energy (in joules and kW·h) generated in 85 minutes.

240 The operating potential difference of a light bulb is 51.1 V and its power rating is 92 W. Calculate the current in the bulb and its resistance.

241 A light bulb has an operating potential difference of 29 V and draws a current of 8.1 A. Calculate its power rating and its resistance.

242 How much does it cost to operate a 43 W light bulb for 16 days if electrical energy costs $0.018 per kW·h?

243 A 29 W lamp is operated for 7 hours a day. How much energy does it take to operate the lamp for 30 days? Express your answer in joules and kilowatt–hours.

244 A resistor is rated at 220 W and its electrical resistance is 1860 Ω.
   a) What is the maximum voltage that can be applied across the resistor?
   b) What is the current at this voltage?

245 Three resistors, 15 Ω, 21 Ω and 25 Ω, are connected in series across a battery that provides 19 V.
   a) Calculate the voltage across each resistor.
   b) Calculate the power dissipated in each resistor.

246 Two resistors, 29 Ω and 39 Ω, are connected in parallel. If the current through the 29 Ω resistor is 3.34 A, find:
    a) The current in the other resistor.
    b) The total power consumed by the two resistors.

247 Find the electric potential drop across a 17 W resistor that draws a current of 3.1 A.

248 A 24 W electric lamp draws a current of 2.7 A. Find the voltage applied and the resistance of the lamp.

249 A 19 kW electric heater operates at 380 V. Find the current, resistance, and energy (in joules and kW·h) generated in 110 minutes.

250 The operating potential difference of a light bulb is 69.8 V and its power rating is 30 W. Calculate the current in the bulb and its resistance.

251 A light bulb has an operating potential difference of 25 V and draws a current of 5 A. Calculate its power rating and its resistance.

252 How much does it cost to operate a 37 W light bulb for 8 days if electrical energy costs $0.042 per kW·h?

253 A 48 W lamp is operated for 12 hours a day. How much energy does it take to operate the lamp for 17 days? Express your answer in joules and kilowatt–hours.

254 A resistor is rated at 135 W and its electrical resistance is 1140 Ω.
    a) What is the maximum voltage that can be applied across the resistor?
    b) What is the current at this voltage?

255 Three resistors, 43 Ω, 56 Ω and 70 Ω, are connected in series across a battery that provides 8 V.
    a) Calculate the voltage across each resistor.
    b) Calculate the power dissipated in each resistor.

256 Two resistors, 18 Ω and 26 Ω, are connected in parallel. If the current through the 18 Ω resistor is 3.33 A, find:
    a) The current in the other resistor.
    b) The total power consumed by the two resistors.

257 Find the electric potential drop across an 81 W resistor that draws a current of 4.9 A.

258 A 66 W electric lamp draws a current of 4.9 A. Find the voltage applied and the resistance of the lamp.

259 A 30 kW electric heater operates at 505 V. Find the current, resistance, and energy (in joules and kW·h) generated in 150 minutes.

260 The operating potential difference of a light bulb is 14.7 V and its power rating is 66 W. Calculate the current in the bulb and its resistance.

261 A light bulb has an operating potential difference of 25 V and draws a current of 6.1 A. Calculate its power rating and its resistance.

262 How much does it cost to operate a 54 W light bulb for 24 days if electrical energy costs $0.12 per kW·h?

263 A 56 W lamp is operated for 13 hours a day. How much energy does it take to operate the lamp for 11 days? Express your answer in joules and kilowatt–hours.

264 A resistor is rated at 45 W and its electrical resistance is 270 Ω.
   a) What is the maximum voltage that can be applied across the resistor?
   b) What is the current at this voltage?

265 Three resistors, 21 Ω, 25 Ω and 30 Ω, are connected in series across a battery that provides 36 V.
   a) Calculate the voltage across each resistor.
   b) Calculate the power dissipated in each resistor.

266 Two resistors, 13 Ω and 26 Ω, are connected in parallel. If the current through the 13 Ω resistor is 4.62 A, find:
   a) The current in the other resistor.
   b) The total power consumed by the two resistors.

267 Find the electric potential drop across a 27 W resistor that draws a current of 4.9 A.

268 A 32 W electric lamp draws a current of 120 mA. Find the voltage applied and the resistance of the lamp.

269 A 38 kW electric heater operates at 370 V. Find the current, resistance, and energy (in joules and kW·h) generated in 55 minutes.

270 The operating potential difference of a light bulb is 5.38 V and its power rating is 14 W. Calculate the current in the bulb and its resistance.

271 A light bulb has an operating potential difference of 33 V and draws a current of 7.8 A. Calculate its power rating and its resistance.

272 How much does it cost to operate a 14 W light bulb for 28 days if electrical energy costs $0.027 per kW·h?

273 A 98 W lamp is operated for 6 hours a day. How much energy does it take to operate the lamp for 6 days? Express your answer in joules and kilowatt–hours.

274 A resistor is rated at 660 W and its electrical resistance is 480 Ω.

　　a) What is the maximum voltage that can be applied across the resistor?

　　b) What is the current at this voltage?

275 Three resistors, 17 Ω, 25 Ω and 39 Ω, are connected in series across a battery that provides 36 V.

　　a) Calculate the voltage across each resistor.

　　b) Calculate the power dissipated in each resistor.

276 Two resistors, 18 Ω and 23 Ω, are connected in parallel. If the current through the 18 Ω resistor is 5 A, find:

　　a) The current in the other resistor.

　　b) The total power consumed by the two resistors.

277 Find the electric potential drop across a 63 W resistor that draws a current of 1.8 A.

278 A 77 W electric lamp draws a current of 3.6 A. Find the voltage applied and the resistance of the lamp.

279 A 39 kW electric heater operates at 580 V. Find the current, resistance, and energy (in joules and kW·h) generated in 50 minutes.

280 The operating potential difference of a light bulb is 48.6 V and its power rating is 18 W. Calculate the current in the bulb and its resistance.

281 A light bulb has an operating potential difference of 37 V and draws a current of 4.7 A. Calculate its power rating and its resistance.

282 How much does it cost to operate a 53 W light bulb for 10 days if electrical energy costs $0.027 per kW·h?

283 A 34 W lamp is operated for 12 hours a day. How much energy does it take to operate the lamp for 12 days? Express your answer in joules and kilowatt–hours.

284 A resistor is rated at 985 W and its electrical resistance is 1810 Ω.

　　a) What is the maximum voltage that can be applied across the resistor?

　　b) What is the current at this voltage?

285 Three resistors, 26 Ω, 40 Ω and 48 Ω, are connected in series across a battery that provides 36 V.

　　a) Calculate the voltage across each resistor.

　　b) Calculate the power dissipated in each resistor.

286 Two resistors, 23 Ω and 33 Ω, are connected in parallel. If the current through the 23 Ω resistor is 3.3 A, find:

    a) The current in the other resistor.

    b) The total power consumed by the two resistors.

287 Find the electric potential drop across a 29 W resistor that draws a current of 2.5 A.

288 A 36 W electric lamp draws a current of 2.9 A. Find the voltage applied and the resistance of the lamp.

289 A 4 kW electric heater operates at 645 V. Find the current, resistance, and energy (in joules and kW·h) generated in 45 minutes.

290 The operating potential difference of a light bulb is 97.8 V and its power rating is 88 W. Calculate the current in the bulb and its resistance.

291 A light bulb has an operating potential difference of 35 V and draws a current of 3.2 A. Calculate its power rating and its resistance.

292 How much does it cost to operate a 73 W light bulb for 18 days if electrical energy costs $0.076 per kW·h?

293 A 52 W lamp is operated for 19 hours a day. How much energy does it take to operate the lamp for 17 days? Express your answer in joules and kilowatt–hours.

294 A resistor is rated at 410 W and its electrical resistance is 1220 Ω.

    a) What is the maximum voltage that can be applied across the resistor?

    b) What is the current at this voltage?

295 Three resistors, 43 Ω, 56 Ω and 66 Ω, are connected in series across a battery that provides 26 V.

    a) Calculate the voltage across each resistor.

    b) Calculate the power dissipated in each resistor.

296 When a capacitor is connected across a 280 V supply the charge is 80.08 μC. Determine:

    a) The capacitance of the capacitor.

    b) The energy stored in the capacitor.

297 An 851 nF capacitor is charged with 76.59 μC. Find the voltage and the energy stored.

298 A capacitor is connected across an 180 V supply. If the energy stored is 4.228 mJ, find:

    a) The capacitance of the capacitor.

    b) The charge on the capacitor.

299 a) Calculate the energy stored in a 695 μF capacitor when charged to 315 V.
b) Find also the charge on the capacitor.

300 A 21 nF capacitor is required to store 3.005 mJ of energy. Find:
    a) Potential difference to which the capacitor must be charged.
    b) The charge on the capacitor.

301 An 145 μF capacitor is charged to a potential difference of 520 V.
    a) Find the charge on the capacitor.
    b) Calculate the energy stored.

302 A capacitor is charged with 100.3 μC. If the energy stored is 11.04 mJ, find:
    a) The voltage applied.
    b) The capacitance of the capacitor.

303 When a capacitor is connected across a 335 V supply the charge is 149.1 mC. Determine:
    a) The capacitance of the capacitor.
    b) The energy stored in the capacitor.

304 A 620 μF capacitor is charged with 12.4 mC. Find the voltage and the energy stored.

305 A capacitor is connected across a 300 V supply. If the energy stored is 24.34 mJ, find:
    a) The capacitance of the capacitor.
    b) The charge on the capacitor.

306 a) Calculate the energy stored in an 121 nF capacitor when charged to 110 V.
b) Find also the charge on the capacitor.

307 A 220 μF capacitor is required to store 4.4 J of energy. Find:
    a) Potential difference to which the capacitor must be charged.
    b) The charge on the capacitor.

308 A 76 nF capacitor is charged to a potential difference of 290 V.
    a) Find the charge on the capacitor.
    b) Calculate the energy stored.

309 A capacitor is charged with 211.8 μC. If the energy stored is 24.36 mJ, find:
    a) The voltage applied.
    b) The capacitance of the capacitor.

310 When a capacitor is connected across a 215 V supply the charge is 41.92 mC. Determine:
    a) The capacitance of the capacitor.
    b) The energy stored in the capacitor.

311 A 711 nF capacitor is charged with 174.2 μC. Find the voltage and the energy stored.

312 A capacitor is connected across a 360 V supply. If the energy stored is 24.04 mJ, find:
   a) The capacitance of the capacitor.
   b) The charge on the capacitor.

313 a) Calculate the energy stored in an 851 nF capacitor when charged to 755 V.
b) Find also the charge on the capacitor.

314 An 176 nF capacitor is required to store 25.66 mJ of energy. Find:
   a) Potential difference to which the capacitor must be charged.
   b) The charge on the capacitor.

315 A 6 nF capacitor is charged to a potential difference of 375 V.
   a) Find the charge on the capacitor.
   b) Calculate the energy stored.

316 A capacitor is charged with 315.7 μC. If the energy stored is 76.57 mJ, find:
   a) The voltage applied.
   b) The capacitance of the capacitor.

317 When a capacitor is connected across an 885 V supply the charge is 8.85 mC. Determine:
   a) The capacitance of the capacitor.
   b) The energy stored in the capacitor.

318 A 51 nF capacitor is charged with 21.17 μC. Find the voltage and the energy stored.

319 A capacitor is connected across a 95 V supply. If the energy stored is 1.494 mJ, find:
   a) The capacitance of the capacitor.
   b) The charge on the capacitor.

320 a) Calculate the energy stored in an 120 μF capacitor when charged to 60 V.
b) Find also the charge on the capacitor.

321 A 626 nF capacitor is required to store 8.521 mJ of energy. Find:
   a) Potential difference to which the capacitor must be charged.
   b) The charge on the capacitor.

322 An 160 μF capacitor is charged to a potential difference of 580 V.
   a) Find the charge on the capacitor.
   b) Calculate the energy stored.

323 A capacitor is charged with 175.5 mC. If the energy stored is 57.04 J, find:
  a) The voltage applied.
  b) The capacitance of the capacitor.

324 When a capacitor is connected across a 395 V supply the charge is 308.1 mC. Determine:
  a) The capacitance of the capacitor.
  b) The energy stored in the capacitor.

325 A 246 nF capacitor is charged with 57.81 μC. Find the voltage and the energy stored.

326 A capacitor is connected across a 480 V supply. If the energy stored is 22 mJ, find:
  a) The capacitance of the capacitor.
  b) The charge on the capacitor.

327 a) Calculate the energy stored in an 845 μF capacitor when charged to 765 V.
b) Find also the charge on the capacitor.

328 A 605 μF capacitor is required to store 218.6 J of energy. Find:
  a) Potential difference to which the capacitor must be charged.
  b) The charge on the capacitor.

329 A 745 μF capacitor is charged to a potential difference of 765 V.
  a) Find the charge on the capacitor.
  b) Calculate the energy stored.

330 A capacitor is charged with 252.8 mC. If the energy stored is 80.9 J, find:
  a) The voltage applied.
  b) The capacitance of the capacitor.

331 When a capacitor is connected across a 650 V supply the charge is 240.5 mC. Determine:
  a) The capacitance of the capacitor.
  b) The energy stored in the capacitor.

332 A 626 nF capacitor is charged with 378.7 μC. Find the voltage and the energy stored.

333 A capacitor is connected across an 875 V supply. If the energy stored is 203.3 mJ, find:
  a) The capacitance of the capacitor.
  b) The charge on the capacitor.

334 a) Calculate the energy stored in an 80 μF capacitor when charged to 75 V.
b) Find also the charge on the capacitor.

335 A 240 μF capacitor is required to store 63.08 J of energy. Find:
   a) Potential difference to which the capacitor must be charged.
   b) The charge on the capacitor.

336 A 306 nF capacitor is charged to a potential difference of 735 V.
   a) Find the charge on the capacitor.
   b) Calculate the energy stored.

337 A capacitor is charged with 82.65 mC. If the energy stored is 11.78 J, find:
   a) The voltage applied.
   b) The capacitance of the capacitor.

338 When a capacitor is connected across an 835 V supply the charge is 513.5 mC. Determine:
   a) The capacitance of the capacitor.
   b) The energy stored in the capacitor.

339 A 5 μF capacitor is charged with 2.65 mC. Find the voltage and the energy stored.

340 A capacitor is connected across a 575 V supply. If the energy stored is 43.81 J, find:
   a) The capacitance of the capacitor.
   b) The charge on the capacitor.

341 **a)** Calculate the energy stored in a 281 nF capacitor when charged to 625 V.
b) Find also the charge on the capacitor.

342 A 586 nF capacitor is required to store 50.46 mJ of energy. Find:
   a) Potential difference to which the capacitor must be charged.
   b) The charge on the capacitor.

343 A 421 nF capacitor is charged to a potential difference of 245 V.
   a) Find the charge on the capacitor.
   b) Calculate the energy stored.

344 A capacitor is charged with 29.16 μC. If the energy stored is 5.249 mJ, find:
   a) The voltage applied.
   b) The capacitance of the capacitor.

345 When a capacitor is connected across a 755 V supply the charge is 423.6 μC. Determine:
   a) The capacitance of the capacitor.
   b) The energy stored in the capacitor.

346 For the circuit shown below, the resistors are $R_1 = 5\ \Omega$, $R_2 = 3\ \Omega$ and $R_3 = 9\ \Omega$. The voltage sources are $\varepsilon_1 = 10\ V$ and $\varepsilon_2 = 32\ V$. Calculate:

      a) Current.

      b) The voltage across R_1.

      c) The voltage across R_2.

      d) Power consumed by R_1.

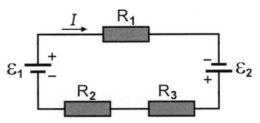

347 For the circuit shown below, the resistors are $R_1 = 5\ \Omega$, $R_2 = 5\ \Omega$ and $R_3 = 3\ \Omega$. The voltage sources are $\varepsilon_1 = 36\ V$ and $\varepsilon_2 = 28\ V$. Calculate:

      a) Current.

      b) The voltage across R_1.

      c) The voltage across R_2.

      d) Power consumed by R_1.

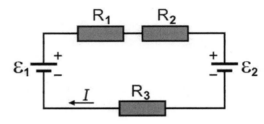

348 For the circuit shown below, the resistors are $R_1 = 5\ \Omega$, $R_2 = 7\ \Omega$ and $R_3 = 5\ \Omega$. The voltage sources are $\varepsilon_1 = 37\ V$ and $\varepsilon_2 = 13\ V$. Calculate:

      a) Current.

      b) The voltage across R_1.

      c) The voltage across R_2.

      d) Power consumed by R_1.

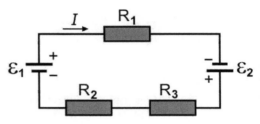

349 For the circuit shown below, the resistors are $R_1 = 8\ \Omega$, $R_2 = 2\ \Omega$ and $R_3 = 10\ \Omega$. The voltage sources are $\varepsilon_1 = 35\ V$ and $\varepsilon_2 = 21\ V$. Calculate:

      a) Current.

      b) The voltage across R_1.

      c) The voltage across R_2.

      d) Power consumed by R_1.

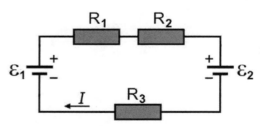

350 For the circuit shown below, the resistors are $R_1 = 11\ \Omega$, $R_2 = 7\ \Omega$ and $R_3 = 5\ \Omega$. The voltage sources are $\varepsilon_1 = 38\ V$ and $\varepsilon_2 = 18\ V$. Calculate:

      a) Current.

      b) The voltage across R_1.

      c) The voltage across R_2.

      d) Power consumed by R_1.

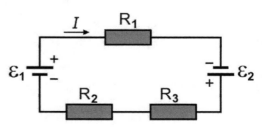

351 For the circuit shown below, the resistors are $R_1 = 5\ \Omega$, $R_2 = 2\ \Omega$ and $R_3 = 10\ \Omega$. The voltage sources are $\varepsilon_1 = 41$ V and $\varepsilon_2 = 7$ V. Calculate:

a) Current.
b) The voltage across R_1.
c) The voltage across R_2.
d) Power consumed by R_1.

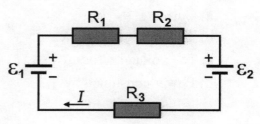

352 For the circuit shown below, the resistors are $R_1 = 11\ \Omega$, $R_2 = 8\ \Omega$ and $R_3 = 4\ \Omega$. The voltage sources are $\varepsilon_1 = 17$ V and $\varepsilon_2 = 17$ V. Calculate:

a) Current.
b) The voltage across R_1.
c) The voltage across R_2.
d) Power consumed by R_1.

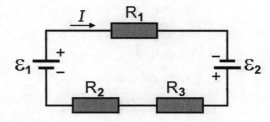

353 For the circuit shown below, the resistors are $R_1 = 2\ \Omega$, $R_2 = 5\ \Omega$ and $R_3 = 1\ \Omega$. The voltage sources are $\varepsilon_1 = 33$ V and $\varepsilon_2 = 30$ V. Calculate:

a) Current.
b) The voltage across R_1.
c) The voltage across R_2.
d) Power consumed by R_1.

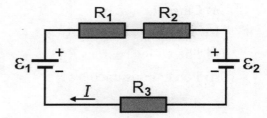

354 For the circuit shown below, the resistors are $R_1 = 4\ \Omega$, $R_2 = 7\ \Omega$ and $R_3 = 5\ \Omega$. The voltage sources are $\varepsilon_1 = 7$ V and $\varepsilon_2 = 28$ V. Calculate:

a) Current.
b) The voltage across R_1.
c) The voltage across R_2.
d) Power consumed by R_1.

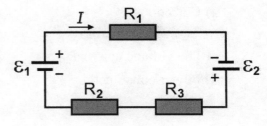

355 For the circuit shown below, the resistors are $R_1 = 5\ \Omega$, $R_2 = 7\ \Omega$ and $R_3 = 6\ \Omega$. The voltage sources are $\varepsilon_1 = 32$ V and $\varepsilon_2 = 17$ V. Calculate:

a) Current.
b) The voltage across R_1.
c) The voltage across R_2.
d) Power consumed by R_1.

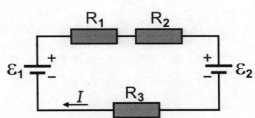

356 For the circuit shown below, the resistors are $R_1 = 11\ \Omega$, $R_2 = 9\ \Omega$ and $R_3 = 2\ \Omega$. The voltage sources are $\varepsilon_1 = 24$ V and $\varepsilon_2 = 16$ V. Calculate:

      a) Current.
      b) The voltage across R_1.
      c) The voltage across R_2.
      d) Power consumed by R_1.

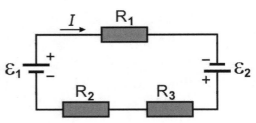

357 For the circuit shown below, the resistors are $R_1 = 3\ \Omega$, $R_2 = 10\ \Omega$ and $R_3 = 3\ \Omega$. The voltage sources are $\varepsilon_1 = 26$ V and $\varepsilon_2 = 11$ V. Calculate:

      a) Current.
      b) The voltage across R_1.
      c) The voltage across R_2.
      d) Power consumed by R_1.

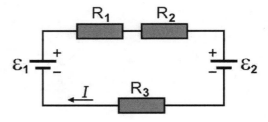

358 For the circuit shown below, the resistors are $R_1 = 8\ \Omega$, $R_2 = 4\ \Omega$ and $R_3 = 6\ \Omega$. The voltage sources are $\varepsilon_1 = 19$ V and $\varepsilon_2 = 40$ V. Calculate:

      a) Current.
      b) The voltage across R_1.
      c) The voltage across R_2.
      d) Power consumed by R_1.

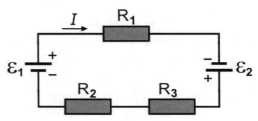

359 For the circuit shown below, the resistors are $R_1 = 10\ \Omega$, $R_2 = 12\ \Omega$ and $R_3 = 9\ \Omega$. The voltage sources are $\varepsilon_1 = 41$ V and $\varepsilon_2 = 13$ V. Calculate:

      a) Current.
      b) The voltage across R_1.
      c) The voltage across R_2.
      d) Power consumed by R_1.

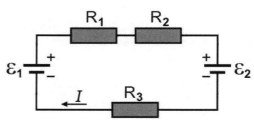

360 For the circuit shown below, the resistors are $R_1 = 6\ \Omega$, $R_2 = 6\ \Omega$ and $R_3 = 3\ \Omega$. The voltage sources are $\varepsilon_1 = 6$ V and $\varepsilon_2 = 12$ V. Calculate:

      a) Current.
      b) The voltage across R_1.
      c) The voltage across R_2.
      d) Power consumed by R_1.

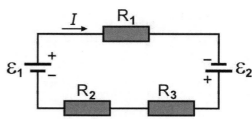

361 For the circuit shown below, the resistors are $R_1 = 3\ \Omega$, $R_2 = 5\ \Omega$ and $R_3 = 5\ \Omega$. The voltage sources are $\varepsilon_1 = 30$ V and $\varepsilon_2 = 6$ V. Calculate:

     a) Current.

     b) The voltage across R_1.

     c) The voltage across R_2.

     d) Power consumed by R_1.

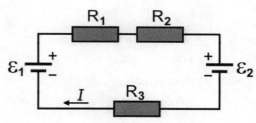

362 For the circuit shown below, the resistors are $R_1 = 3\ \Omega$, $R_2 = 6\ \Omega$ and $R_3 = 7\ \Omega$. The voltage sources are $\varepsilon_1 = 40$ V and $\varepsilon_2 = 29$ V. Calculate:

     a) Current.

     b) The voltage across R_1.

     c) The voltage across R_2.

     d) Power consumed by R_1.

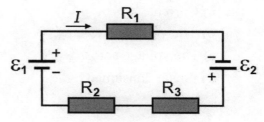

363 For the circuit shown below, the resistors are $R_1 = 3\ \Omega$, $R_2 = 12\ \Omega$ and $R_3 = 11\ \Omega$. The voltage sources are $\varepsilon_1 = 44$ V and $\varepsilon_2 = 28$ V. Calculate:

     a) Current.

     b) The voltage across R_1.

     c) The voltage across R_2.

     d) Power consumed by R_1.

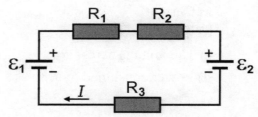

364 For the circuit shown below, the resistors are $R_1 = 7\ \Omega$, $R_2 = 2\ \Omega$ and $R_3 = 5\ \Omega$. The voltage sources are $\varepsilon_1 = 7$ V and $\varepsilon_2 = 30$ V. Calculate:

     a) Current.

     b) The voltage across R_1.

     c) The voltage across R_2.

     d) Power consumed by R_1.

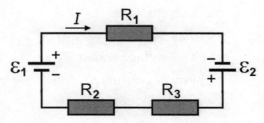

365 For the circuit shown below, the resistors are $R_1 = 4\ \Omega$, $R_2 = 10\ \Omega$ and $R_3 = 12\ \Omega$. The voltage sources are $\varepsilon_1 = 43$ V and $\varepsilon_2 = 23$ V. Calculate:

     a) Current.

     b) The voltage across R_1.

     c) The voltage across R_2.

     d) Power consumed by R_1.

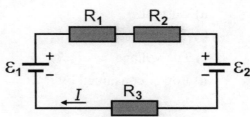

366 For the circuit shown below, the resistors are $R_1 = 9\,\Omega$, $R_2 = 9\,\Omega$ and $R_3 = 11\,\Omega$. The voltage sources are $\varepsilon_1 = 43$ V and $\varepsilon_2 = 45$ V. Calculate:

      a) Current.
      b) The voltage across R_1.
      c) The voltage across R_2.
      d) Power consumed by R_1.

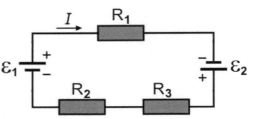

367 For the circuit shown below, the resistors are $R_1 = 3\,\Omega$, $R_2 = 5\,\Omega$ and $R_3 = 10\,\Omega$. The voltage sources are $\varepsilon_1 = 27$ V and $\varepsilon_2 = 24$ V. Calculate:

      a) Current.
      b) The voltage across R_1.
      c) The voltage across R_2.
      d) Power consumed by R_1.

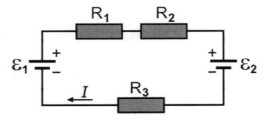

368 For the circuit shown below, the resistors are $R_1 = 9\,\Omega$, $R_2 = 2\,\Omega$ and $R_3 = 3\,\Omega$. The voltage sources are $\varepsilon_1 = 25$ V and $\varepsilon_2 = 14$ V. Calculate:

      a) Current.
      b) The voltage across R_1.
      c) The voltage across R_2.
      d) Power consumed by R_1.

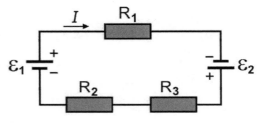

369 For the circuit shown below, the resistors are $R_1 = 11\,\Omega$, $R_2 = 6\,\Omega$ and $R_3 = 1\,\Omega$. The voltage sources are $\varepsilon_1 = 14$ V and $\varepsilon_2 = 7$ V. Calculate:

      a) Current.
      b) The voltage across R_1.
      c) The voltage across R_2.
      d) Power consumed by R_1.

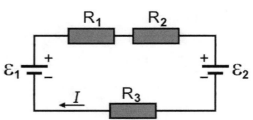

370 For the circuit shown below, the resistors are $R_1 = 11\,\Omega$, $R_2 = 3\,\Omega$ and $R_3 = 7\,\Omega$. The voltage sources are $\varepsilon_1 = 9$ V and $\varepsilon_2 = 11$ V. Calculate:

      a) Current.
      b) The voltage across R_1.
      c) The voltage across R_2.
      d) Power consumed by R_1.

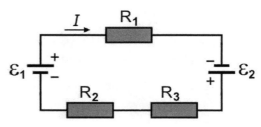

371 For the circuit shown below, the resistors are $R_1 = 11\ \Omega$, $R_2 = 7\ \Omega$ and $R_3 = 1\ \Omega$. The voltage sources are $\varepsilon_1 = 33$ V and $\varepsilon_2 = 6$ V. Calculate:

    a) Current.
    b) The voltage across R_1.
    c) The voltage across R_2.
    d) Power consumed by R_1.

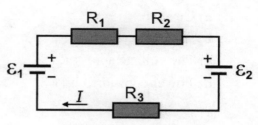

372 For the circuit shown below, the resistors are $R_1 = 12\ \Omega$, $R_2 = 7\ \Omega$ and $R_3 = 12\ \Omega$. The voltage sources are $\varepsilon_1 = 23$ V and $\varepsilon_2 = 31$ V. Calculate:

    a) Current.
    b) The voltage across R_1.
    c) The voltage across R_2.
    d) Power consumed by R_1.

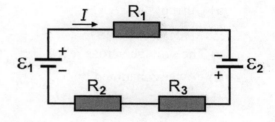

373 For the circuit shown below, the resistors are $R_1 = 5\ \Omega$, $R_2 = 10\ \Omega$ and $R_3 = 6\ \Omega$. The voltage sources are $\varepsilon_1 = 28$ V and $\varepsilon_2 = 22$ V. Calculate:

    a) Current.
    b) The voltage across R_1.
    c) The voltage across R_2.
    d) Power consumed by R_1.

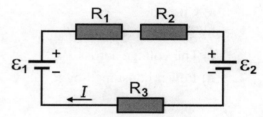

374 For the circuit shown below, the resistors are $R_1 = 11\ \Omega$, $R_2 = 8\ \Omega$ and $R_3 = 3\ \Omega$. The voltage sources are $\varepsilon_1 = 15$ V and $\varepsilon_2 = 31$ V. Calculate:

    a) Current.
    b) The voltage across R_1.
    c) The voltage across R_2.
    d) Power consumed by R_1.

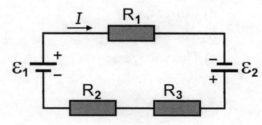

375 For the circuit shown below, the resistors are $R_1 = 6\ \Omega$, $R_2 = 5\ \Omega$ and $R_3 = 9\ \Omega$. The voltage sources are $\varepsilon_1 = 17$ V and $\varepsilon_2 = 7$ V. Calculate:

    a) Current.
    b) The voltage across R_1.
    c) The voltage across R_2.
    d) Power consumed by R_1.

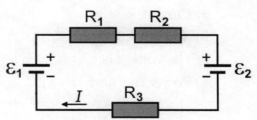

376 For the circuit shown below, the resistors are $R_1 = 3\ \Omega$, $R_2 = 5\ \Omega$ and $R_3 = 6\ \Omega$. The voltage sources are $\varepsilon_1 = 18$ V and $\varepsilon_2 = 44$ V. Calculate:

    a) Current.
    b) The voltage across R_1.
    c) The voltage across R_2.
    d) Power consumed by R_1.

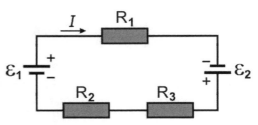

377 For the circuit shown below, the resistors are $R_1 = 3\ \Omega$, $R_2 = 4\ \Omega$ and $R_3 = 4\ \Omega$. The voltage sources are $\varepsilon_1 = 37$ V and $\varepsilon_2 = 19$ V. Calculate:

    a) Current.
    b) The voltage across R_1.
    c) The voltage across R_2.
    d) Power consumed by R_1.

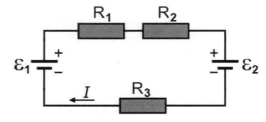

378 For the circuit shown below, the resistors are $R_1 = 7\ \Omega$, $R_2 = 7\ \Omega$ and $R_3 = 12\ \Omega$. The voltage sources are $\varepsilon_1 = 13$ V and $\varepsilon_2 = 11$ V. Calculate:

    a) Current.
    b) The voltage across R_1.
    c) The voltage across R_2.
    d) Power consumed by R_1.

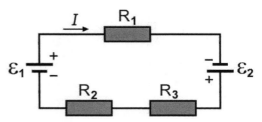

379 For the circuit shown below, the resistors are $R_1 = 4\ \Omega$, $R_2 = 5\ \Omega$ and $R_3 = 2\ \Omega$. The voltage sources are $\varepsilon_1 = 44$ V and $\varepsilon_2 = 18$ V. Calculate:

    a) Current.
    b) The voltage across R_1.
    c) The voltage across R_2.
    d) Power consumed by R_1.

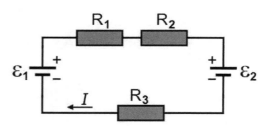

380 For the circuit shown below, the resistors are $R_1 = 4\ \Omega$, $R_2 = 12\ \Omega$ and $R_3 = 12\ \Omega$. The voltage sources are $\varepsilon_1 = 24$ V and $\varepsilon_2 = 34$ V. Calculate:

    a) Current.
    b) The voltage across R_1.
    c) The voltage across R_2.
    d) Power consumed by R_1.

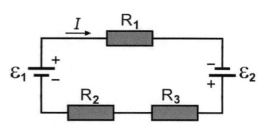

381 For the circuit shown below, the resistors are $R_1 = 9\ \Omega$, $R_2 = 11\ \Omega$ and $R_3 = 12\ \Omega$. The voltage sources are $\varepsilon_1 = 34$ V and $\varepsilon_2 = 14$ V. Calculate:

    a) Current.
    b) The voltage across R_1.
    c) The voltage across R_2.
    d) Power consumed by R_1.

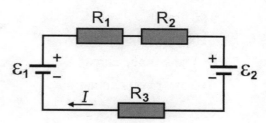

382 For the circuit shown below, the resistors are $R_1 = 8\ \Omega$, $R_2 = 2\ \Omega$ and $R_3 = 3\ \Omega$. The voltage sources are $\varepsilon_1 = 35$ V and $\varepsilon_2 = 30$ V. Calculate:

    a) Current.
    b) The voltage across R_1.
    c) The voltage across R_2.
    d) Power consumed by R_1.

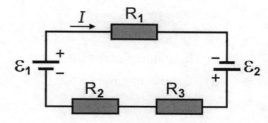

383 For the circuit shown below, the resistors are $R_1 = 6\ \Omega$, $R_2 = 5\ \Omega$ and $R_3 = 11\ \Omega$. The voltage sources are $\varepsilon_1 = 20$ V and $\varepsilon_2 = 9$ V. Calculate:

    a) Current.
    b) The voltage across R_1.
    c) The voltage across R_2.
    d) Power consumed by R_1.

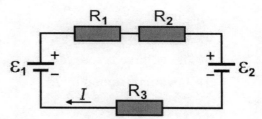

384 For the circuit shown below, the resistors are $R_1 = 2\ \Omega$, $R_2 = 8\ \Omega$ and $R_3 = 11\ \Omega$. The voltage sources are $\varepsilon_1 = 33$ V and $\varepsilon_2 = 15$ V. Calculate:

    a) Current.
    b) The voltage across R_1.
    c) The voltage across R_2.
    d) Power consumed by R_1.

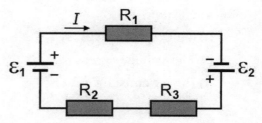

385 For the circuit shown below, the resistors are $R_1 = 11\ \Omega$, $R_2 = 4\ \Omega$ and $R_3 = 11\ \Omega$. The voltage sources are $\varepsilon_1 = 31$ V and $\varepsilon_2 = 27$ V. Calculate:

    a) Current.
    b) The voltage across R_1.
    c) The voltage across R_2.
    d) Power consumed by R_1.

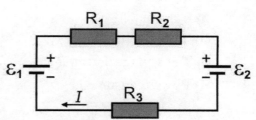

386 For the circuit shown below, the resistors are $R_1 = 8\,\Omega$, $R_2 = 7\,\Omega$ and $R_3 = 5\,\Omega$. The voltage sources are $\varepsilon_1 = 6$ V and $\varepsilon_2 = 31$ V. Calculate:

        a) Current.
        b) The voltage across R_1.
        c) The voltage across R_2.
        d) Power consumed by R_1.

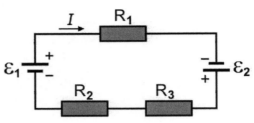

387 For the circuit shown below, the resistors are $R_1 = 3\,\Omega$, $R_2 = 4\,\Omega$ and $R_3 = 7\,\Omega$. The voltage sources are $\varepsilon_1 = 38$ V and $\varepsilon_2 = 20$ V. Calculate:

        a) Current.
        b) The voltage across R_1.
        c) The voltage across R_2.
        d) Power consumed by R_1.

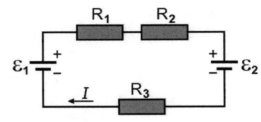

388 For the circuit shown below, the resistors are $R_1 = 2\,\Omega$, $R_2 = 6\,\Omega$ and $R_3 = 7\,\Omega$. The voltage sources are $\varepsilon_1 = 10$ V and $\varepsilon_2 = 12$ V. Calculate:

        a) Current.
        b) The voltage across R_1.
        c) The voltage across R_2.
        d) Power consumed by R_1.

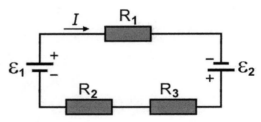

389 For the circuit shown below, the resistors are $R_1 = 4\,\Omega$, $R_2 = 4\,\Omega$ and $R_3 = 1\,\Omega$. The voltage sources are $\varepsilon_1 = 44$ V and $\varepsilon_2 = 26$ V. Calculate:

        a) Current.
        b) The voltage across R_1.
        c) The voltage across R_2.
        d) Power consumed by R_1.

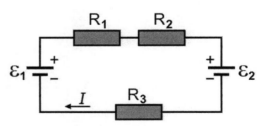

390 For the circuit shown below, the resistors are $R_1 = 12\,\Omega$, $R_2 = 8\,\Omega$ and $R_3 = 10\,\Omega$. The voltage sources are $\varepsilon_1 = 13$ V and $\varepsilon_2 = 23$ V. Calculate:

        a) Current.
        b) The voltage across R_1.
        c) The voltage across R_2.
        d) Power consumed by R_1.

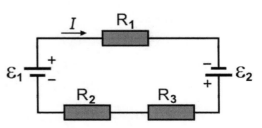

391 For the circuit shown below, the resistors are $R_1 = 4\,\Omega$, $R_2 = 10\,\Omega$ and $R_3 = 7\,\Omega$. The voltage sources are $\varepsilon_1 = 37$ V and $\varepsilon_2 = 5$ V. Calculate:

   a) Current.
   b) The voltage across R_1.
   c) The voltage across R_2.
   d) Power consumed by R_1.

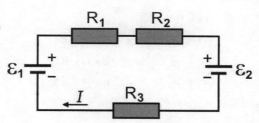

392 For the circuit shown below, the resistors are $R_1 = 6\,\Omega$, $R_2 = 4\,\Omega$ and $R_3 = 8\,\Omega$. The voltage sources are $\varepsilon_1 = 11$ V and $\varepsilon_2 = 9$ V. Calculate:

   a) Current.
   b) The voltage across R_1.
   c) The voltage across R_2.
   d) Power consumed by R_1.

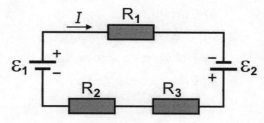

393 For the circuit shown below, the resistors are $R_1 = 11\,\Omega$, $R_2 = 6\,\Omega$ and $R_3 = 9\,\Omega$. The voltage sources are $\varepsilon_1 = 29$ V and $\varepsilon_2 = 26$ V. Calculate:

   a) Current.
   b) The voltage across R_1.
   c) The voltage across R_2.
   d) Power consumed by R_1.

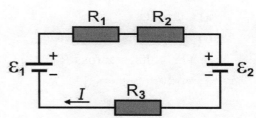

394 For the circuit shown below, the resistors are $R_1 = 10\,\Omega$, $R_2 = 3\,\Omega$ and $R_3 = 8\,\Omega$. The voltage sources are $\varepsilon_1 = 40$ V and $\varepsilon_2 = 19$ V. Calculate:

   a) Current.
   b) The voltage across R_1.
   c) The voltage across R_2.
   d) Power consumed by R_1.

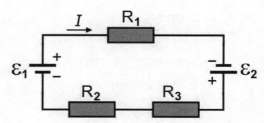

395 For the circuit shown below, the resistors are $R_1 = 11\,\Omega$, $R_2 = 7\,\Omega$ and $R_3 = 9\,\Omega$. The voltage sources are $\varepsilon_1 = 41$ V and $\varepsilon_2 = 13$ V. Calculate:

   a) Current.
   b) The voltage across R_1.
   c) The voltage across R_2.
   d) Power consumed by R_1.

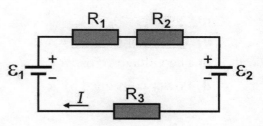

396 For the circuit shown below, the resistors are $R_1 = 8\,\Omega$, $R_2 = 3\,\Omega$ and $R_3 = 9\,\Omega$. The voltage sources are $\varepsilon_1 = 41$ V and $\varepsilon_2 = 8$ V. Find:

> **a)** Currents.
> **b)** The voltage across R_1.
> **c)** The voltage across R_2.
> **d)** Power consumed by R_1.

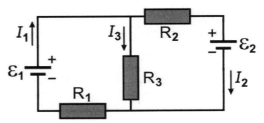

397 For the circuit shown below, the resistors are $R_1 = 4\,\Omega$, $R_2 = 7\,\Omega$ and $R_3 = 2\,\Omega$. The voltage sources are $\varepsilon_1 = 36$ V and $\varepsilon_2 = 20$ V. Find:

> **a)** Currents.
> **b)** The voltage across R_1.
> **c)** The voltage across R_2.
> **d)** Power consumed by R_1.

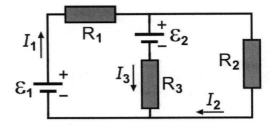

398 For the circuit shown below, the resistors are $R_1 = 6\,\Omega$, $R_2 = 9\,\Omega$ and $R_3 = 1\,\Omega$. The voltage sources are $\varepsilon_1 = 11$ V and $\varepsilon_2 = 9$ V. Find:

> **a)** Currents.
> **b)** The voltage across R_1.
> **c)** The voltage across R_2.
> **d)** Power consumed by R_1.

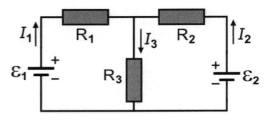

399 For the circuit shown below, the resistors are $R_1 = 4\,\Omega$, $R_2 = 6\,\Omega$, $R_3 = 6\,\Omega$ and $R_4 = 5\,\Omega$. The voltage sources are $\varepsilon_1 = 27$ V and $\varepsilon_2 = 28$ V. Find:

> **a)** Currents.
> **b)** The voltage across R_1.
> **c)** The voltage across R_2.
> **d)** Power consumed by R_1.

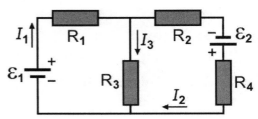

400 For the circuit shown below, the resistors are $R_1 = 2\,\Omega$, $R_2 = 3\,\Omega$, $R_3 = 9\,\Omega$, $R_4 = 9\,\Omega$ and $R_5 = 2\,\Omega$. The voltage sources are $\varepsilon_1 = 35$ V, $\varepsilon_2 = 16$ V and $\varepsilon_3 = 20$ V. Find:

> **a)** Currents.
> **b)** The voltage across R_1.
> **c)** The voltage across R_2.
> **d)** Power consumed by R_1.

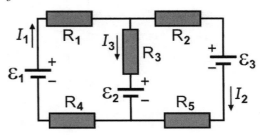

401 For the circuit shown below, the resistors are $R_1 = 11\ \Omega$, $R_2 = 11\ \Omega$ and $R_3 = 11\ \Omega$. The voltage sources are $\varepsilon_1 = 38$ V and $\varepsilon_2 = 7$ V. Find:

    a) Currents.
    b) The voltage across R_1.
    c) The voltage across R_2.
    d) Power consumed by R_1.

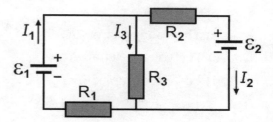

402 For the circuit shown below, the resistors are $R_1 = 9\ \Omega$, $R_2 = 9\ \Omega$ and $R_3 = 2\ \Omega$. The voltage sources are $\varepsilon_1 = 39$ V and $\varepsilon_2 = 11$ V. Find:

    a) Currents.
    b) The voltage across R_1.
    c) The voltage across R_2.
    d) Power consumed by R_1.

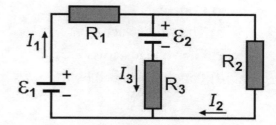

403 For the circuit shown below, the resistors are $R_1 = 5\ \Omega$, $R_2 = 6\ \Omega$ and $R_3 = 4\ \Omega$. The voltage sources are $\varepsilon_1 = 15$ V and $\varepsilon_2 = 27$ V. Find:

    a) Currents.
    b) The voltage across R_1.
    c) The voltage across R_2.
    d) Power consumed by R_1.

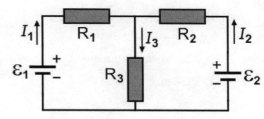

404 For the circuit shown below, the resistors are $R_1 = 6\ \Omega$, $R_2 = 9\ \Omega$, $R_3 = 1\ \Omega$ and $R_4 = 1\ \Omega$. The voltage sources are $\varepsilon_1 = 35$ V and $\varepsilon_2 = 42$ V. Find:

    a) Currents.
    b) The voltage across R_1.
    c) The voltage across R_2.
    d) Power consumed by R_1.

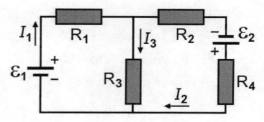

405 For the circuit shown below, the resistors are $R_1 = 3\ \Omega$, $R_2 = 7\ \Omega$, $R_3 = 9\ \Omega$, $R_4 = 8\ \Omega$ and $R_5 = 12\ \Omega$. The voltage sources are $\varepsilon_1 = 37$ V, $\varepsilon_2 = 27$ V and $\varepsilon_3 = 25$ V. Find:

    a) Currents.
    b) The voltage across R_1.
    c) The voltage across R_2.
    d) Power consumed by R_1.

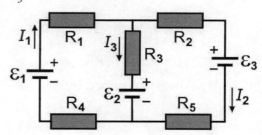

406 For the circuit shown below, the resistors are $R_1 = 5\ \Omega$, $R_2 = 3\ \Omega$ and $R_3 = 8\ \Omega$. The voltage sources are $\varepsilon_1 = 33$ V and $\varepsilon_2 = 6$ V. Find:

      a) Currents.
      b) The voltage across R_1.
      c) The voltage across R_2.
      d) Power consumed by R_1.

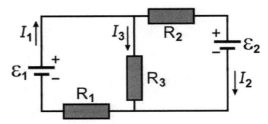

407 For the circuit shown below, the resistors are $R_1 = 4\ \Omega$, $R_2 = 8\ \Omega$ and $R_3 = 3\ \Omega$. The voltage sources are $\varepsilon_1 = 25$ V and $\varepsilon_2 = 11$ V. Find:

      a) Currents.
      b) The voltage across R_1.
      c) The voltage across R_2.
      d) Power consumed by R_1.

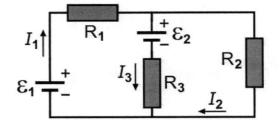

408 For the circuit shown below, the resistors are $R_1 = 8\ \Omega$, $R_2 = 12\ \Omega$ and $R_3 = 12\ \Omega$. The voltage sources are $\varepsilon_1 = 37$ V and $\varepsilon_2 = 30$ V. Find:

      a) Currents.
      b) The voltage across R_1.
      c) The voltage across R_2.
      d) Power consumed by R_1.

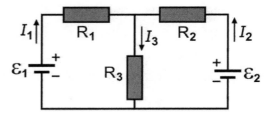

409 For the circuit shown below, the resistors are $R_1 = 12\ \Omega$, $R_2 = 11\ \Omega$, $R_3 = 5\ \Omega$ and $R_4 = 6\ \Omega$. The voltage sources are $\varepsilon_1 = 33$ V and $\varepsilon_2 = 23$ V. Find:

      a) Currents.
      b) The voltage across R_1.
      c) The voltage across R_2.
      d) Power consumed by R_1.

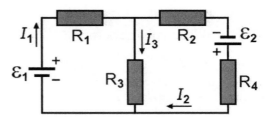

410 For the circuit shown below, the resistors are $R_1 = 4\ \Omega$, $R_2 = 7\ \Omega$, $R_3 = 2\ \Omega$, $R_4 = 4\ \Omega$ and $R_5 = 2\ \Omega$. The voltage sources are $\varepsilon_1 = 36$ V, $\varepsilon_2 = 8$ V and $\varepsilon_3 = 9$ V. Find:

      a) Currents.
      b) The voltage across R_1.
      c) The voltage across R_2.
      d) Power consumed by R_1.

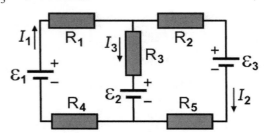

411 For the circuit shown below, the resistors are $R_1 = 8\ \Omega$, $R_2 = 8\ \Omega$ and $R_3 = 9\ \Omega$. The voltage sources are $\varepsilon_1 = 26$ V and $\varepsilon_2 = 6$ V. Find:

    a) Currents.
    b) The voltage across R_1.
    c) The voltage across R_2.
    d) Power consumed by R_1.

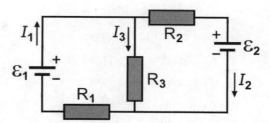

412 For the circuit shown below, the resistors are $R_1 = 5\ \Omega$, $R_2 = 6\ \Omega$ and $R_3 = 12\ \Omega$. The voltage sources are $\varepsilon_1 = 29$ V and $\varepsilon_2 = 14$ V. Find:

    a) Currents.
    b) The voltage across R_1.
    c) The voltage across R_2.
    d) Power consumed by R_1.

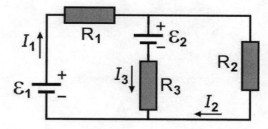

413 For the circuit shown below, the resistors are $R_1 = 5\ \Omega$, $R_2 = 4\ \Omega$ and $R_3 = 12\ \Omega$. The voltage sources are $\varepsilon_1 = 14$ V and $\varepsilon_2 = 14$ V. Find:

    a) Currents.
    b) The voltage across R_1.
    c) The voltage across R_2.
    d) Power consumed by R_1.

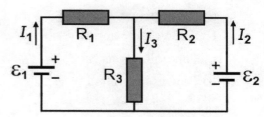

414 For the circuit shown below, the resistors are $R_1 = 11\ \Omega$, $R_2 = 3\ \Omega$, $R_3 = 8\ \Omega$ and $R_4 = 3\ \Omega$. The voltage sources are $\varepsilon_1 = 27$ V and $\varepsilon_2 = 9$ V. Find:

    a) Currents.
    b) The voltage across R_1.
    c) The voltage across R_2.
    d) Power consumed by R_1.

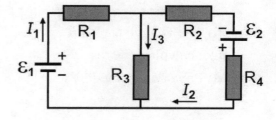

415 For the circuit shown below, the resistors are $R_1 = 9\ \Omega$, $R_2 = 7\ \Omega$, $R_3 = 7\ \Omega$, $R_4 = 3\ \Omega$ and $R_5 = 3\ \Omega$. The voltage sources are $\varepsilon_1 = 35$ V, $\varepsilon_2 = 6$ V and $\varepsilon_3 = 12$ V. Find:

    a) Currents.
    b) The voltage across R_1.
    c) The voltage across R_2.
    d) Power consumed by R_1.

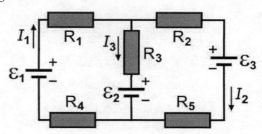

416 For the circuit shown below, the resistors are $R_1 = 5\,\Omega$, $R_2 = 8\,\Omega$ and $R_3 = 6\,\Omega$. The voltage sources are $\varepsilon_1 = 34$ V and $\varepsilon_2 = 9$ V. Find:

    a) Currents.
    b) The voltage across R_1.
    c) The voltage across R_2.
    d) Power consumed by R_1.

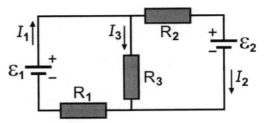

417 For the circuit shown below, the resistors are $R_1 = 6\,\Omega$, $R_2 = 9\,\Omega$ and $R_3 = 2\,\Omega$. The voltage sources are $\varepsilon_1 = 30$ V and $\varepsilon_2 = 7$ V. Find:

    a) Currents.
    b) The voltage across R_1.
    c) The voltage across R_2.
    d) Power consumed by R_1.

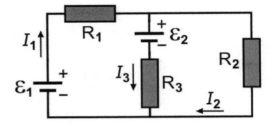

418 For the circuit shown below, the resistors are $R_1 = 10\,\Omega$, $R_2 = 9\,\Omega$ and $R_3 = 4\,\Omega$. The voltage sources are $\varepsilon_1 = 12$ V and $\varepsilon_2 = 29$ V. Find:

    a) Currents.
    b) The voltage across R_1.
    c) The voltage across R_2.
    d) Power consumed by R_1.

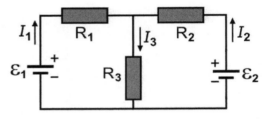

419 For the circuit shown below, the resistors are $R_1 = 4\,\Omega$, $R_2 = 4\,\Omega$, $R_3 = 2\,\Omega$ and $R_4 = 10\,\Omega$. The voltage sources are $\varepsilon_1 = 40$ V and $\varepsilon_2 = 30$ V. Find:

    a) Currents.
    b) The voltage across R_1.
    c) The voltage across R_2.
    d) Power consumed by R_1.

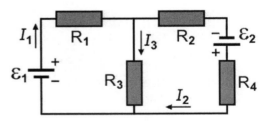

420 For the circuit shown below, the resistors are $R_1 = 3\,\Omega$, $R_2 = 11\,\Omega$, $R_3 = 6\,\Omega$, $R_4 = 2\,\Omega$ and $R_5 = 3\,\Omega$. The voltage sources are $\varepsilon_1 = 42$ V, $\varepsilon_2 = 14$ V and $\varepsilon_3 = 15$ V. Find:

    a) Currents.
    b) The voltage across R_1.
    c) The voltage across R_2.
    d) Power consumed by R_1.

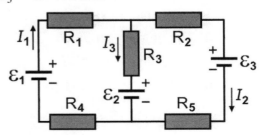

421 For the circuit shown below, the resistors are $R_1 = 4\,\Omega$, $R_2 = 7\,\Omega$ and $R_3 = 4\,\Omega$. The voltage sources are $\varepsilon_1 = 41$ V and $\varepsilon_2 = 13$ V. Find:

    a) Currents.

    b) The voltage across R_1.

    c) The voltage across R_2.

    d) Power consumed by R_1.

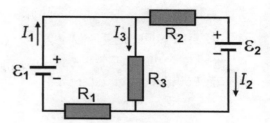

422 For the circuit shown below, the resistors are $R_1 = 3\,\Omega$, $R_2 = 5\,\Omega$ and $R_3 = 8\,\Omega$. The voltage sources are $\varepsilon_1 = 24$ V and $\varepsilon_2 = 7$ V. Find:

    a) Currents.

    b) The voltage across R_1.

    c) The voltage across R_2.

    d) Power consumed by R_1.

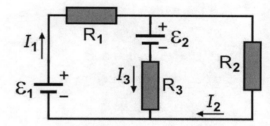

423 For the circuit shown below, the resistors are $R_1 = 11\,\Omega$, $R_2 = 6\,\Omega$ and $R_3 = 9\,\Omega$. The voltage sources are $\varepsilon_1 = 37$ V and $\varepsilon_2 = 18$ V. Find:

    a) Currents.

    b) The voltage across R_1.

    c) The voltage across R_2.

    d) Power consumed by R_1.

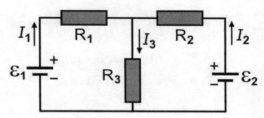

424 For the circuit shown below, the resistors are $R_1 = 8\,\Omega$, $R_2 = 11\,\Omega$, $R_3 = 12\,\Omega$ and $R_4 = 10\,\Omega$. The voltage sources are $\varepsilon_1 = 29$ V and $\varepsilon_2 = 17$ V. Find:

    a) Currents.

    b) The voltage across R_1.

    c) The voltage across R_2.

    d) Power consumed by R_1.

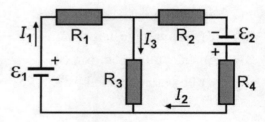

425 For the circuit shown below, the resistors are $R_1 = 10\,\Omega$, $R_2 = 6\,\Omega$, $R_3 = 7\,\Omega$, $R_4 = 2\,\Omega$ and $R_5 = 4\,\Omega$. The voltage sources are $\varepsilon_1 = 19$ V, $\varepsilon_2 = 11$ V and $\varepsilon_3 = 9$ V. Find:

    a) Currents.

    b) The voltage across R_1.

    c) The voltage across R_2.

    d) Power consumed by R_1.

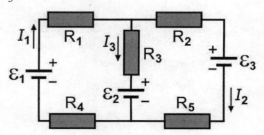

426 For the circuit shown below, the resistors are $R_1 = 11\ \Omega$, $R_2 = 4\ \Omega$ and $R_3 = 7\ \Omega$. The voltage sources are $\varepsilon_1 = 43$ V and $\varepsilon_2 = 11$ V. Find:

      a) Currents.
      b) The voltage across R_1.
      c) The voltage across R_2.
      d) Power consumed by R_1.

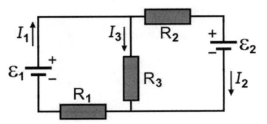

427 For the circuit shown below, the resistors are $R_1 = 3\ \Omega$, $R_2 = 8\ \Omega$ and $R_3 = 11\ \Omega$. The voltage sources are $\varepsilon_1 = 32$ V and $\varepsilon_2 = 19$ V. Find:

      a) Currents.
      b) The voltage across R_1.
      c) The voltage across R_2.
      d) Power consumed by R_1.

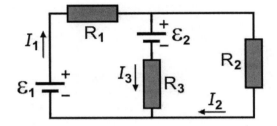

428 For the circuit shown below, the resistors are $R_1 = 2\ \Omega$, $R_2 = 7\ \Omega$ and $R_3 = 4\ \Omega$. The voltage sources are $\varepsilon_1 = 11$ V and $\varepsilon_2 = 29$ V. Find:

      a) Currents.
      b) The voltage across R_1.
      c) The voltage across R_2.
      d) Power consumed by R_1.

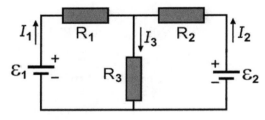

429 For the circuit shown below, the resistors are $R_1 = 8\ \Omega$, $R_2 = 4\ \Omega$, $R_3 = 2\ \Omega$ and $R_4 = 12\ \Omega$. The voltage sources are $\varepsilon_1 = 26$ V and $\varepsilon_2 = 26$ V. Find:

      a) Currents.
      b) The voltage across R_1.
      c) The voltage across R_2.
      d) Power consumed by R_1.

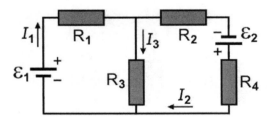

430 For the circuit shown below, the resistors are $R_1 = 3\ \Omega$, $R_2 = 4\ \Omega$, $R_3 = 6\ \Omega$, $R_4 = 6\ \Omega$ and $R_5 = 5\ \Omega$. The voltage sources are $\varepsilon_1 = 43$ V, $\varepsilon_2 = 17$ V and $\varepsilon_3 = 24$ V. Find:

      a) Currents.
      b) The voltage across R_1.
      c) The voltage across R_2.
      d) Power consumed by R_1.

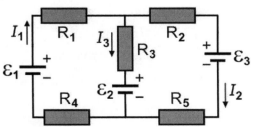

431 For the circuit shown below, the resistors are $R_1 = 10\,\Omega$, $R_2 = 5\,\Omega$ and $R_3 = 10\,\Omega$. The voltage sources are $\varepsilon_1 = 37$ V and $\varepsilon_2 = 8$ V. Find:

    a) Currents.
    b) The voltage across R_1.
    c) The voltage across R_2.
    d) Power consumed by R_1.

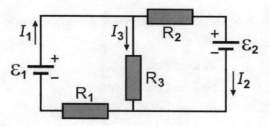

432 For the circuit shown below, the resistors are $R_1 = 4\,\Omega$, $R_2 = 9\,\Omega$ and $R_3 = 10\,\Omega$. The voltage sources are $\varepsilon_1 = 29$ V and $\varepsilon_2 = 16$ V. Find:

    a) Currents.
    b) The voltage across R_1.
    c) The voltage across R_2.
    d) Power consumed by R_1.

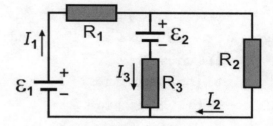

433 For the circuit shown below, the resistors are $R_1 = 6\,\Omega$, $R_2 = 2\,\Omega$ and $R_3 = 7\,\Omega$. The voltage sources are $\varepsilon_1 = 35$ V and $\varepsilon_2 = 43$ V. Find:

    a) Currents.
    b) The voltage across R_1.
    c) The voltage across R_2.
    d) Power consumed by R_1.

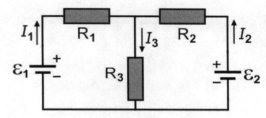

434 For the circuit shown below, the resistors are $R_1 = 5\,\Omega$, $R_2 = 6\,\Omega$, $R_3 = 9\,\Omega$ and $R_4 = 3\,\Omega$. The voltage sources are $\varepsilon_1 = 39$ V and $\varepsilon_2 = 31$ V. Find:

    a) Currents.
    b) The voltage across R_1.
    c) The voltage across R_2.
    d) Power consumed by R_1.

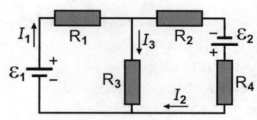

435 For the circuit shown below, the resistors are $R_1 = 4\,\Omega$, $R_2 = 12\,\Omega$, $R_3 = 2\,\Omega$, $R_4 = 6\,\Omega$ and $R_5 = 6\,\Omega$. The voltage sources are $\varepsilon_1 = 29$ V, $\varepsilon_2 = 8$ V and $\varepsilon_3 = 9$ V. Find:

    a) Currents.
    b) The voltage across R_1.
    c) The voltage across R_2.
    d) Power consumed by R_1.

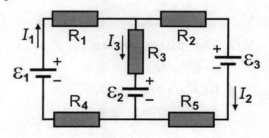

436 For the circuit shown below, the resistors are $R_1 = 10\ \Omega$, $R_2 = 6\ \Omega$ and $R_3 = 4\ \Omega$. The voltage sources are $\varepsilon_1 = 37$ V and $\varepsilon_2 = 6$ V. Find:

    a) Currents.
    b) The voltage across R_1.
    c) The voltage across R_2.
    d) Power consumed by R_1.

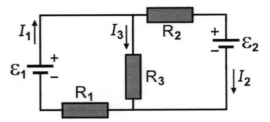

437 For the circuit shown below, the resistors are $R_1 = 8\ \Omega$, $R_2 = 11\ \Omega$ and $R_3 = 2\ \Omega$. The voltage sources are $\varepsilon_1 = 37$ V and $\varepsilon_2 = 19$ V. Find:

    a) Currents.
    b) The voltage across R_1.
    c) The voltage across R_2.
    d) Power consumed by R_1.

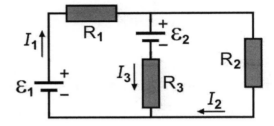

438 For the circuit shown below, the resistors are $R_1 = 5\ \Omega$, $R_2 = 8\ \Omega$ and $R_3 = 4\ \Omega$. The voltage sources are $\varepsilon_1 = 32$ V and $\varepsilon_2 = 24$ V. Find:

    a) Currents.
    b) The voltage across R_1.
    c) The voltage across R_2.
    d) Power consumed by R_1.

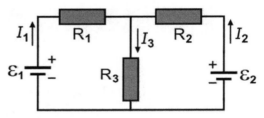

439 For the circuit shown below, the resistors are $R_1 = 9\ \Omega$, $R_2 = 11\ \Omega$, $R_3 = 10\ \Omega$ and $R_4 = 11\ \Omega$. The voltage sources are $\varepsilon_1 = 37$ V and $\varepsilon_2 = 15$ V. Find:

    a) Currents.
    b) The voltage across R_1.
    c) The voltage across R_2.
    d) Power consumed by R_1.

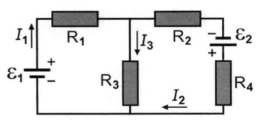

440 For the circuit shown below, the resistors are $R_1 = 3\ \Omega$, $R_2 = 4\ \Omega$, $R_3 = 1\ \Omega$, $R_4 = 4\ \Omega$ and $R_5 = 9\ \Omega$. The voltage sources are $\varepsilon_1 = 38$ V, $\varepsilon_2 = 19$ V and $\varepsilon_3 = 14$ V. Find:

    a) Currents.
    b) The voltage across R_1.
    c) The voltage across R_2.
    d) Power consumed by R_1.

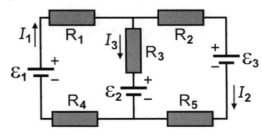

441 For the circuit shown below, the resistors are $R_1 = 6\ \Omega$, $R_2 = 11\ \Omega$ and $R_3 = 12\ \Omega$. The voltage sources are $\varepsilon_1 = 25$ V and $\varepsilon_2 = 14$ V. Find:

    a) Currents.
    b) The voltage across R_1.
    c) The voltage across R_2.
    d) Power consumed by R_1.

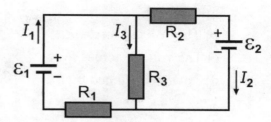

442 For the circuit shown below, the resistors are $R_1 = 4\ \Omega$, $R_2 = 6\ \Omega$ and $R_3 = 12\ \Omega$. The voltage sources are $\varepsilon_1 = 34$ V and $\varepsilon_2 = 18$ V. Find:

    a) Currents.
    b) The voltage across R_1.
    c) The voltage across R_2.
    d) Power consumed by R_1.

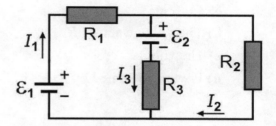

443 For the circuit shown below, the resistors are $R_1 = 6\ \Omega$, $R_2 = 11\ \Omega$ and $R_3 = 1\ \Omega$. The voltage sources are $\varepsilon_1 = 40$ V and $\varepsilon_2 = 8$ V. Find:

    a) Currents.
    b) The voltage across R_1.
    c) The voltage across R_2.
    d) Power consumed by R_1.

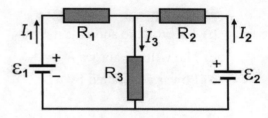

444 For the circuit shown below, the resistors are $R_1 = 5\ \Omega$, $R_2 = 8\ \Omega$, $R_3 = 8\ \Omega$ and $R_4 = 4\ \Omega$. The voltage sources are $\varepsilon_1 = 27$ V and $\varepsilon_2 = 20$ V. Find:

    a) Currents.
    b) The voltage across R_1.
    c) The voltage across R_2.
    d) Power consumed by R_1.

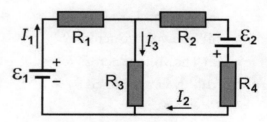

445 For the circuit shown below, the resistors are $R_1 = 7\ \Omega$, $R_2 = 8\ \Omega$, $R_3 = 7\ \Omega$, $R_4 = 7\ \Omega$ and $R_5 = 10\ \Omega$. The voltage sources are $\varepsilon_1 = 39$ V, $\varepsilon_2 = 29$ V and $\varepsilon_3 = 24$ V. Find:

    a) Currents.
    b) The voltage across R_1.
    c) The voltage across R_2.
    d) Power consumed by R_1.

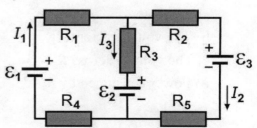

446 For the circuit shown below, the resistors are $R_1 = 12\ \Omega$, $R_2 = 4\ \Omega$ and $R_3 = 8\ \Omega$. The voltage sources are $\varepsilon_1 = 44$ V and $\varepsilon_2 = 9$ V. Find:

        a) Currents.
        b) The voltage across R_1.
        c) The voltage across R_2.
        d) Power consumed by R_1.

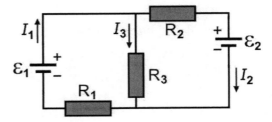

447 For the circuit shown below, the resistors are $R_1 = 5\ \Omega$, $R_2 = 11\ \Omega$ and $R_3 = 9\ \Omega$. The voltage sources are $\varepsilon_1 = 32$ V and $\varepsilon_2 = 13$ V. Find:

        a) Currents.
        b) The voltage across R_1.
        c) The voltage across R_2.
        d) Power consumed by R_1.

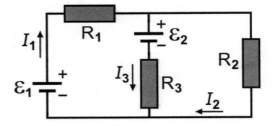

448 For the circuit shown below, the resistors are $R_1 = 5\ \Omega$, $R_2 = 3\ \Omega$ and $R_3 = 11\ \Omega$. The voltage sources are $\varepsilon_1 = 18$ V and $\varepsilon_2 = 14$ V. Find:

        a) Currents.
        b) The voltage across R_1.
        c) The voltage across R_2.
        d) Power consumed by R_1.

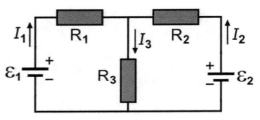

449 For the circuit shown below, the resistors are $R_1 = 5\ \Omega$, $R_2 = 12\ \Omega$, $R_3 = 9\ \Omega$ and $R_4 = 5\ \Omega$. The voltage sources are $\varepsilon_1 = 37$ V and $\varepsilon_2 = 28$ V. Find:

        a) Currents.
        b) The voltage across R_1.
        c) The voltage across R_2.
        d) Power consumed by R_1.

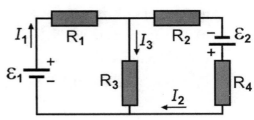

450 For the circuit shown below, the resistors are $R_1 = 10\ \Omega$, $R_2 = 2\ \Omega$, $R_3 = 11\ \Omega$, $R_4 = 3\ \Omega$ and $R_5 = 4\ \Omega$. The voltage sources are $\varepsilon_1 = 43$ V, $\varepsilon_2 = 19$ V and $\varepsilon_3 = 14$ V. Find:

        a) Currents.
        b) The voltage across R_1.
        c) The voltage across R_2.
        d) Power consumed by R_1.

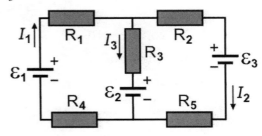

451 For the circuit shown below, the resistors are $R_1 = 3\ \Omega$, $R_2 = 3\ \Omega$, $R_3 = 2\ \Omega$, $R_4 = 7\ \Omega$, $R_5 = 10\ \Omega$ and $R_6 = 5\ \Omega$. The voltage sources are $\varepsilon_1 = 27$ V and $\varepsilon_2 = 16$ V. Find:

  a) Currents.
  b) The voltage across R_1.
  c) The voltage across R_2.
  d) Power consumed by R_1.

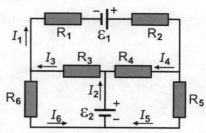

452 For the circuit shown below, the resistors are $R_1 = 4\ \Omega$, $R_2 = 11\ \Omega$, $R_3 = 6\ \Omega$, $R_4 = 6\ \Omega$, $R_5 = 12\ \Omega$ and $R_6 = 7\ \Omega$. The voltage sources are $\varepsilon_1 = 45$ V, $\varepsilon_2 = 25$ V and $\varepsilon_3 = 45$ V. Find:

  a) Currents.
  b) The voltage across R_1.
  c) The voltage across R_2.
  d) Power consumed by R_1.

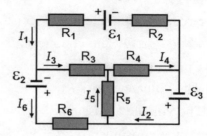

453 For the circuit shown below, the resistors are $R_1 = 10\ \Omega$, $R_2 = 8\ \Omega$, $R_3 = 2\ \Omega$, $R_4 = 7\ \Omega$, $R_5 = 10\ \Omega$ and $R_6 = 1\ \Omega$. The voltage sources are $\varepsilon_1 = 26$ V and $\varepsilon_2 = 8$ V. Find:

  a) Currents.
  b) The voltage across R_1.
  c) The voltage across R_2.
  d) Power consumed by R_1.

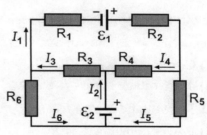

454 For the circuit shown below, the resistors are $R_1 = 3\ \Omega$, $R_2 = 11\ \Omega$, $R_3 = 11\ \Omega$, $R_4 = 10\ \Omega$, $R_5 = 5\ \Omega$ and $R_6 = 11\ \Omega$. The voltage sources are $\varepsilon_1 = 44$ V, $\varepsilon_2 = 8$ V and $\varepsilon_3 = 31$ V. Find:

  a) Currents.
  b) The voltage across R_1.
  c) The voltage across R_2.
  d) Power consumed by R_1.

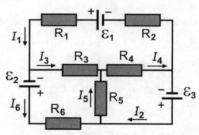

455 For the circuit shown below, the resistors are $R_1 = 10\ \Omega$, $R_2 = 7\ \Omega$, $R_3 = 2\ \Omega$, $R_4 = 3\ \Omega$, $R_5 = 11\ \Omega$ and $R_6 = 2\ \Omega$. The voltage sources are $\varepsilon_1 = 25$ V and $\varepsilon_2 = 9$ V. Find:

  a) Currents.
  b) The voltage across R_1.
  c) The voltage across R_2.
  d) Power consumed by R_1.

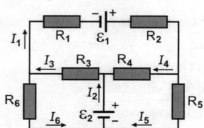

456 For the circuit shown below, the resistors are $R_1 = 10\ \Omega$, $R_2 = 4\ \Omega$, $R_3 = 3\ \Omega$, $R_4 = 4\ \Omega$, $R_5 = 8\ \Omega$ and $R_6 = 6\ \Omega$. The voltage sources are $\varepsilon_1 = 26$ V, $\varepsilon_2 = 26$ V and $\varepsilon_3 = 37$ V. Find:

      a) Currents.

      b) The voltage across R_1.

      c) The voltage across R_2.

      d) Power consumed by R_1.

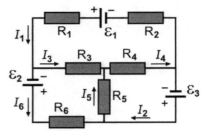

457 For the circuit shown below, the resistors are $R_1 = 6\ \Omega$, $R_2 = 3\ \Omega$, $R_3 = 3\ \Omega$, $R_4 = 7\ \Omega$, $R_5 = 12\ \Omega$ and $R_6 = 2\ \Omega$. The voltage sources are $\varepsilon_1 = 40$ V and $\varepsilon_2 = 21$ V. Find:

      a) Currents.

      b) The voltage across R_1.

      c) The voltage across R_2.

      d) Power consumed by R_1.

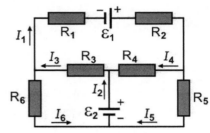

458 For the circuit shown below, the resistors are $R_1 = 10\ \Omega$, $R_2 = 12\ \Omega$, $R_3 = 9\ \Omega$, $R_4 = 10\ \Omega$, $R_5 = 9\ \Omega$ and $R_6 = 4\ \Omega$. The voltage sources are $\varepsilon_1 = 32$ V, $\varepsilon_2 = 14$ V and $\varepsilon_3 = 31$ V. Find:

      a) Currents.

      b) The voltage across R_1.

      c) The voltage across R_2.

      d) Power consumed by R_1.

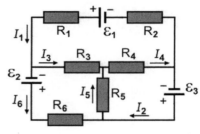

459 For the circuit shown below, the resistors are $R_1 = 5\ \Omega$, $R_2 = 8\ \Omega$, $R_3 = 3\ \Omega$, $R_4 = 4\ \Omega$, $R_5 = 7\ \Omega$ and $R_6 = 5\ \Omega$. The voltage sources are $\varepsilon_1 = 40$ V and $\varepsilon_2 = 14$ V. Find:

      a) Currents.

      b) The voltage across R_1.

      c) The voltage across R_2.

      d) Power consumed by R_1.

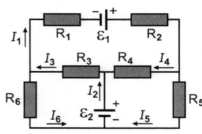

460 For the circuit shown below, the resistors are $R_1 = 9\ \Omega$, $R_2 = 4\ \Omega$, $R_3 = 11\ \Omega$, $R_4 = 3\ \Omega$, $R_5 = 7\ \Omega$ and $R_6 = 8\ \Omega$. The voltage sources are $\varepsilon_1 = 23$ V, $\varepsilon_2 = 29$ V and $\varepsilon_3 = 36$ V. Find:

      a) Currents.

      b) The voltage across R_1.

      c) The voltage across R_2.

      d) Power consumed by R_1.

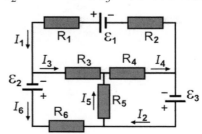

461 For the circuit shown below, the resistors are $R_1 = 7\ \Omega$, $R_2 = 12\ \Omega$, $R_3 = 2\ \Omega$, $R_4 = 4\ \Omega$, $R_5 = 10\ \Omega$ and $R_6 = 6\ \Omega$. The voltage sources are $\varepsilon_1 = 38$ V and $\varepsilon_2 = 6$ V. Find:

    a) Currents.

    b) The voltage across R_1.

    c) The voltage across R_2.

    d) Power consumed by R_1.

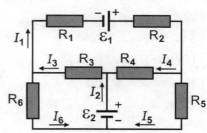

462 For the circuit shown below, the resistors are $R_1 = 4\ \Omega$, $R_2 = 3\ \Omega$, $R_3 = 5\ \Omega$, $R_4 = 12\ \Omega$, $R_5 = 11\ \Omega$ and $R_6 = 10\ \Omega$. The voltage sources are $\varepsilon_1 = 27$ V, $\varepsilon_2 = 15$ V and $\varepsilon_3 = 30$ V. Find:

    a) Currents.

    b) The voltage across R_1.

    c) The voltage across R_2.

    d) Power consumed by R_1.

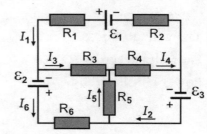

463 For the circuit shown below, the resistors are $R_1 = 4\ \Omega$, $R_2 = 3\ \Omega$, $R_3 = 2\ \Omega$, $R_4 = 4\ \Omega$, $R_5 = 9\ \Omega$ and $R_6 = 8\ \Omega$. The voltage sources are $\varepsilon_1 = 31$ V and $\varepsilon_2 = 11$ V. Find:

    a) Currents.

    b) The voltage across R_1.

    c) The voltage across R_2.

    d) Power consumed by R_1.

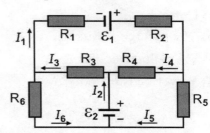

464 For the circuit shown below, the resistors are $R_1 = 2\ \Omega$, $R_2 = 5\ \Omega$, $R_3 = 10\ \Omega$, $R_4 = 3\ \Omega$, $R_5 = 2\ \Omega$ and $R_6 = 5\ \Omega$. The voltage sources are $\varepsilon_1 = 23$ V, $\varepsilon_2 = 11$ V and $\varepsilon_3 = 22$ V. Find:

    a) Currents.

    b) The voltage across R_1.

    c) The voltage across R_2.

    d) Power consumed by R_1.

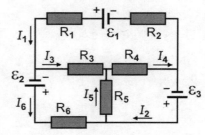

465 For the circuit shown below, the resistors are $R_1 = 10\ \Omega$, $R_2 = 8\ \Omega$, $R_3 = 3\ \Omega$, $R_4 = 7\ \Omega$, $R_5 = 10\ \Omega$ and $R_6 = 3\ \Omega$. The voltage sources are $\varepsilon_1 = 24$ V and $\varepsilon_2 = 6$ V. Find:

    a) Currents.

    b) The voltage across R_1.

    c) The voltage across R_2.

    d) Power consumed by R_1.

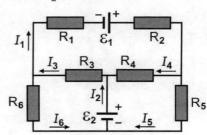

466 For the circuit shown below, the resistors are $R_1 = 6\ \Omega$, $R_2 = 4\ \Omega$, $R_3 = 11\ \Omega$, $R_4 = 3\ \Omega$, $R_5 = 7\ \Omega$ and $R_6 = 10\ \Omega$. The voltage sources are $\varepsilon_1 = 42$ V, $\varepsilon_2 = 15$ V and $\varepsilon_3 = 24$ V. Find:

    a) Currents.
    b) The voltage across R_1.
    c) The voltage across R_2.
    d) Power consumed by R_1.

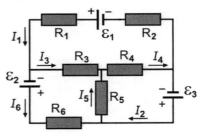

467 For the circuit shown below, the resistors are $R_1 = 10\ \Omega$, $R_2 = 4\ \Omega$, $R_3 = 2\ \Omega$, $R_4 = 7\ \Omega$, $R_5 = 11\ \Omega$ and $R_6 = 4\ \Omega$. The voltage sources are $\varepsilon_1 = 41$ V and $\varepsilon_2 = 10$ V. Find:

    a) Currents.
    b) The voltage across R_1.
    c) The voltage across R_2.
    d) Power consumed by R_1.

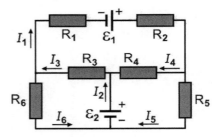

468 For the circuit shown below, the resistors are $R_1 = 9\ \Omega$, $R_2 = 11\ \Omega$, $R_3 = 11\ \Omega$, $R_4 = 8\ \Omega$, $R_5 = 5\ \Omega$ and $R_6 = 8\ \Omega$. The voltage sources are $\varepsilon_1 = 41$ V, $\varepsilon_2 = 6$ V and $\varepsilon_3 = 23$ V. Find:

    a) Currents.
    b) The voltage across R_1.
    c) The voltage across R_2.
    d) Power consumed by R_1.

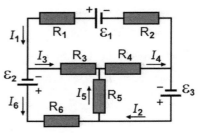

469 For the circuit shown below, the resistors are $R_1 = 7\ \Omega$, $R_2 = 10\ \Omega$, $R_3 = 1\ \Omega$, $R_4 = 8\ \Omega$, $R_5 = 5\ \Omega$ and $R_6 = 4\ \Omega$. The voltage sources are $\varepsilon_1 = 34$ V and $\varepsilon_2 = 5$ V. Find:

    a) Currents.
    b) The voltage across R_1.
    c) The voltage across R_2.
    d) Power consumed by R_1.

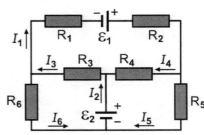

470 For the circuit shown below, the resistors are $R_1 = 3\ \Omega$, $R_2 = 7\ \Omega$, $R_3 = 3\ \Omega$, $R_4 = 2\ \Omega$, $R_5 = 5\ \Omega$ and $R_6 = 7\ \Omega$. The voltage sources are $\varepsilon_1 = 40$ V, $\varepsilon_2 = 40$ V and $\varepsilon_3 = 33$ V. Find:

    a) Currents.
    b) The voltage across R_1.
    c) The voltage across R_2.
    d) Power consumed by R_1.

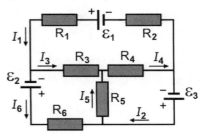

471 For the circuit shown below, the resistors are $R_1 = 10 \, \Omega$, $R_2 = 11 \, \Omega$, $R_3 = 2 \, \Omega$, $R_4 = 11 \, \Omega$, $R_5 = 11 \, \Omega$ and $R_6 = 1 \, \Omega$. The voltage sources are $\varepsilon_1 = 45$ V and $\varepsilon_2 = 14$ V. Find:

    a) Currents.

    b) The voltage across R_1.

    c) The voltage across R_2.

    d) Power consumed by R_1.

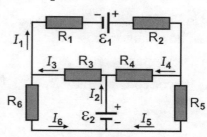

472 For the circuit shown below, the resistors are $R_1 = 10 \, \Omega$, $R_2 = 2 \, \Omega$, $R_3 = 8 \, \Omega$, $R_4 = 3 \, \Omega$, $R_5 = 7 \, \Omega$ and $R_6 = 4 \, \Omega$. The voltage sources are $\varepsilon_1 = 39$ V, $\varepsilon_2 = 25$ V and $\varepsilon_3 = 44$ V. Find:

    a) Currents.

    b) The voltage across R_1.

    c) The voltage across R_2.

    d) Power consumed by R_1.

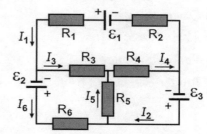

473 For the circuit shown below, the resistors are $R_1 = 6 \, \Omega$, $R_2 = 4 \, \Omega$, $R_3 = 1 \, \Omega$, $R_4 = 6 \, \Omega$, $R_5 = 9 \, \Omega$ and $R_6 = 11 \, \Omega$. The voltage sources are $\varepsilon_1 = 30$ V and $\varepsilon_2 = 11$ V. Find:

    a) Currents.

    b) The voltage across R_1.

    c) The voltage across R_2.

    d) Power consumed by R_1.

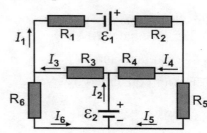

474 For the circuit shown below, the resistors are $R_1 = 6 \, \Omega$, $R_2 = 7 \, \Omega$, $R_3 = 11 \, \Omega$, $R_4 = 6 \, \Omega$, $R_5 = 9 \, \Omega$ and $R_6 = 7 \, \Omega$. The voltage sources are $\varepsilon_1 = 33$ V, $\varepsilon_2 = 29$ V and $\varepsilon_3 = 41$ V. Find:

    a) Currents.

    b) The voltage across R_1.

    c) The voltage across R_2.

    d) Power consumed by R_1.

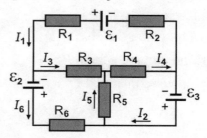

475 For the circuit shown below, the resistors are $R_1 = 8 \, \Omega$, $R_2 = 10 \, \Omega$, $R_3 = 2 \, \Omega$, $R_4 = 11 \, \Omega$, $R_5 = 6 \, \Omega$ and $R_6 = 12 \, \Omega$. The voltage sources are $\varepsilon_1 = 44$ V and $\varepsilon_2 = 8$ V. Find:

    a) Currents.

    b) The voltage across R_1.

    c) The voltage across R_2.

    d) Power consumed by R_1.

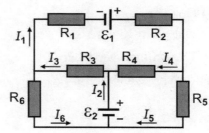

476 For the circuit shown below, the resistors are $R_1 = 3\,\Omega$, $R_2 = 2\,\Omega$, $R_3 = 7\,\Omega$, $R_4 = 7\,\Omega$, $R_5 = 11\,\Omega$ and $R_6 = 6\,\Omega$. The voltage sources are $\varepsilon_1 = 39$ V, $\varepsilon_2 = 27$ V and $\varepsilon_3 = 48$ V. Find:

    a) Currents.
    b) The voltage across R_1.
    c) The voltage across R_2.
    d) Power consumed by R_1.

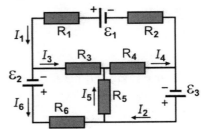

477 For the circuit shown below, the resistors are $R_1 = 8\,\Omega$, $R_2 = 4\,\Omega$, $R_3 = 2\,\Omega$, $R_4 = 1\,\Omega$, $R_5 = 10\,\Omega$ and $R_6 = 1\,\Omega$. The voltage sources are $\varepsilon_1 = 34$ V and $\varepsilon_2 = 13$ V. Find:

    a) Currents.
    b) The voltage across R_1.
    c) The voltage across R_2.
    d) Power consumed by R_1.

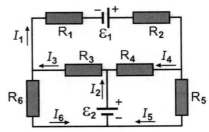

478 For the circuit shown below, the resistors are $R_1 = 9\,\Omega$, $R_2 = 5\,\Omega$, $R_3 = 8\,\Omega$, $R_4 = 5\,\Omega$, $R_5 = 10\,\Omega$ and $R_6 = 10\,\Omega$. The voltage sources are $\varepsilon_1 = 40$ V, $\varepsilon_2 = 18$ V and $\varepsilon_3 = 30$ V. Find:

    a) Currents.
    b) The voltage across R_1.
    c) The voltage across R_2.
    d) Power consumed by R_1.

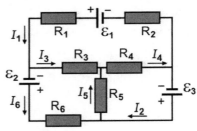

479 For the circuit shown below, the resistors are $R_1 = 10\,\Omega$, $R_2 = 10\,\Omega$, $R_3 = 2\,\Omega$, $R_4 = 11\,\Omega$, $R_5 = 12\,\Omega$ and $R_6 = 10\,\Omega$. The voltage sources are $\varepsilon_1 = 44$ V and $\varepsilon_2 = 11$ V. Find:

    a) Currents.
    b) The voltage across R_1.
    c) The voltage across R_2.
    d) Power consumed by R_1.

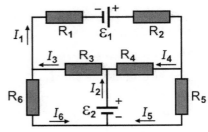

480 For the circuit shown below, the resistors are $R_1 = 3\,\Omega$, $R_2 = 5\,\Omega$, $R_3 = 6\,\Omega$, $R_4 = 11\,\Omega$, $R_5 = 5\,\Omega$ and $R_6 = 4\,\Omega$. The voltage sources are $\varepsilon_1 = 38$ V, $\varepsilon_2 = 14$ V and $\varepsilon_3 = 39$ V. Find:

    a) Currents.
    b) The voltage across R_1.
    c) The voltage across R_2.
    d) Power consumed by R_1.

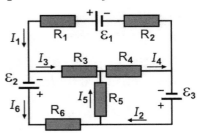

481 For the circuit shown below, the resistors are $R_1 = 4\,\Omega$, $R_2 = 7\,\Omega$, $R_3 = 2\,\Omega$, $R_4 = 9\,\Omega$, $R_5 = 9\,\Omega$ and $R_6 = 4\,\Omega$. The voltage sources are $\varepsilon_1 = 41$ V and $\varepsilon_2 = 17$ V. Find:

    a) Currents.

    b) The voltage across R_1.

    c) The voltage across R_2.

    d) Power consumed by R_1.

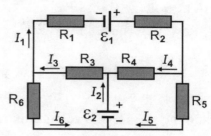

482 For the circuit shown below, the resistors are $R_1 = 3\,\Omega$, $R_2 = 5\,\Omega$, $R_3 = 8\,\Omega$, $R_4 = 4\,\Omega$, $R_5 = 9\,\Omega$ and $R_6 = 5\,\Omega$. The voltage sources are $\varepsilon_1 = 33$ V, $\varepsilon_2 = 26$ V and $\varepsilon_3 = 38$ V. Find:

    a) Currents.

    b) The voltage across R_1.

    c) The voltage across R_2.

    d) Power consumed by R_1.

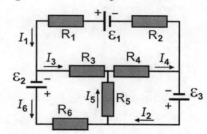

483 For the circuit shown below, the resistors are $R_1 = 8\,\Omega$, $R_2 = 11\,\Omega$, $R_3 = 2\,\Omega$, $R_4 = 3\,\Omega$, $R_5 = 12\,\Omega$ and $R_6 = 9\,\Omega$. The voltage sources are $\varepsilon_1 = 33$ V and $\varepsilon_2 = 12$ V. Find:

    a) Currents.

    b) The voltage across R_1.

    c) The voltage across R_2.

    d) Power consumed by R_1.

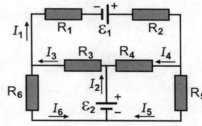

484 For the circuit shown below, the resistors are $R_1 = 5\,\Omega$, $R_2 = 6\,\Omega$, $R_3 = 6\,\Omega$, $R_4 = 3\,\Omega$, $R_5 = 10\,\Omega$ and $R_6 = 11\,\Omega$. The voltage sources are $\varepsilon_1 = 43$ V, $\varepsilon_2 = 42$ V and $\varepsilon_3 = 44$ V. Find:

    a) Currents.

    b) The voltage across R_1.

    c) The voltage across R_2.

    d) Power consumed by R_1.

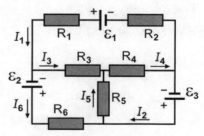

485 For the circuit shown below, the resistors are $R_1 = 6\,\Omega$, $R_2 = 10\,\Omega$, $R_3 = 1\,\Omega$, $R_4 = 8\,\Omega$, $R_5 = 6\,\Omega$ and $R_6 = 11\,\Omega$. The voltage sources are $\varepsilon_1 = 39$ V and $\varepsilon_2 = 9$ V. Find:

    a) Currents.

    b) The voltage across R_1.

    c) The voltage across R_2.

    d) Power consumed by R_1.

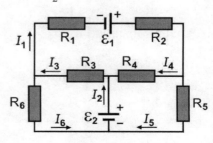

486 For the circuit shown below, the resistors are $R_1 = 6\ \Omega$, $R_2 = 10\ \Omega$, $R_3 = 9\ \Omega$, $R_4 = 5\ \Omega$, $R_5 = 3\ \Omega$ and $R_6 = 7\ \Omega$. The voltage sources are $\varepsilon_1 = 45$ V, $\varepsilon_2 = 20$ V and $\varepsilon_3 = 46$ V. Find:

    a) Currents.
    b) The voltage across R_1.
    c) The voltage across R_2.
    d) Power consumed by R_1.

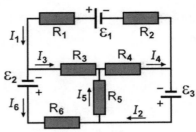

487 For the circuit shown below, the resistors are $R_1 = 6\ \Omega$, $R_2 = 9\ \Omega$, $R_3 = 4\ \Omega$, $R_4 = 8\ \Omega$, $R_5 = 12\ \Omega$ and $R_6 = 2\ \Omega$. The voltage sources are $\varepsilon_1 = 40$ V and $\varepsilon_2 = 15$ V. Find:

    a) Currents.
    b) The voltage across R_1.
    c) The voltage across R_2.
    d) Power consumed by R_1.

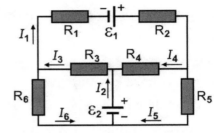

488 For the circuit shown below, the resistors are $R_1 = 4\ \Omega$, $R_2 = 10\ \Omega$, $R_3 = 10\ \Omega$, $R_4 = 1\ \Omega$, $R_5 = 11\ \Omega$ and $R_6 = 11\ \Omega$. The voltage sources are $\varepsilon_1 = 15$ V, $\varepsilon_2 = 25$ V and $\varepsilon_3 = 26$ V. Find:

    a) Currents.
    b) The voltage across R_1.
    c) The voltage across R_2.
    d) Power consumed by R_1.

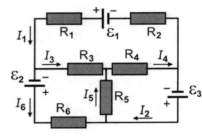

489 For the circuit shown below, the resistors are $R_1 = 7\ \Omega$, $R_2 = 6\ \Omega$, $R_3 = 2\ \Omega$, $R_4 = 8\ \Omega$, $R_5 = 8\ \Omega$ and $R_6 = 8\ \Omega$. The voltage sources are $\varepsilon_1 = 37$ V and $\varepsilon_2 = 12$ V. Find:

    a) Currents.
    b) The voltage across R_1.
    c) The voltage across R_2.
    d) Power consumed by R_1.

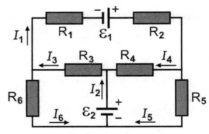

490 For the circuit shown below, the resistors are $R_1 = 3\ \Omega$, $R_2 = 4\ \Omega$, $R_3 = 6\ \Omega$, $R_4 = 2\ \Omega$, $R_5 = 8\ \Omega$ and $R_6 = 3\ \Omega$. The voltage sources are $\varepsilon_1 = 14$ V, $\varepsilon_2 = 17$ V and $\varepsilon_3 = 26$ V. Find:

    a) Currents.
    b) The voltage across R_1.
    c) The voltage across R_2.
    d) Power consumed by R_1.

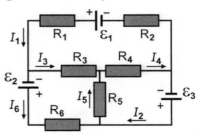

491 For the circuit shown below, the resistors are $R_1 = 5\,\Omega$, $R_2 = 11\,\Omega$, $R_3 = 1\,\Omega$, $R_4 = 11\,\Omega$, $R_5 = 10\,\Omega$ and $R_6 = 11\,\Omega$. The voltage sources are $\varepsilon_1 = 40$ V and $\varepsilon_2 = 17$ V. Find:

    a) Currents.

    b) The voltage across R_1.

    c) The voltage across R_2.

    d) Power consumed by R_1.

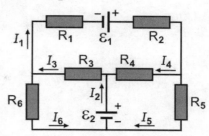

492 For the circuit shown below, the resistors are $R_1 = 11\,\Omega$, $R_2 = 10\,\Omega$, $R_3 = 2\,\Omega$, $R_4 = 1\,\Omega$, $R_5 = 2\,\Omega$ and $R_6 = 10\,\Omega$. The voltage sources are $\varepsilon_1 = 30$ V, $\varepsilon_2 = 18$ V and $\varepsilon_3 = 14$ V. Find:

    a) Currents.

    b) The voltage across R_1.

    c) The voltage across R_2.

    d) Power consumed by R_1.

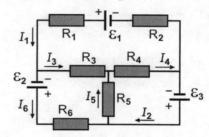

493 For the circuit shown below, the resistors are $R_1 = 11\,\Omega$, $R_2 = 8\,\Omega$, $R_3 = 4\,\Omega$, $R_4 = 10\,\Omega$, $R_5 = 10\,\Omega$ and $R_6 = 4\,\Omega$. The voltage sources are $\varepsilon_1 = 34$ V and $\varepsilon_2 = 9$ V. Find:

    a) Currents.

    b) The voltage across R_1.

    c) The voltage across R_2.

    d) Power consumed by R_1.

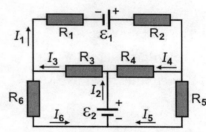

494 For the circuit shown below, the resistors are $R_1 = 8\,\Omega$, $R_2 = 5\,\Omega$, $R_3 = 6\,\Omega$, $R_4 = 6\,\Omega$, $R_5 = 5\,\Omega$ and $R_6 = 10\,\Omega$. The voltage sources are $\varepsilon_1 = 22$ V, $\varepsilon_2 = 12$ V and $\varepsilon_3 = 20$ V. Find:

    a) Currents.

    b) The voltage across R_1.

    c) The voltage across R_2.

    d) Power consumed by R_1.

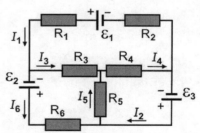

495 For the circuit shown below, the resistors are $R_1 = 3\,\Omega$, $R_2 = 11\,\Omega$, $R_3 = 3\,\Omega$, $R_4 = 9\,\Omega$, $R_5 = 7\,\Omega$ and $R_6 = 4\,\Omega$. The voltage sources are $\varepsilon_1 = 36$ V and $\varepsilon_2 = 10$ V. Find:

    a) Currents.

    b) The voltage across R_1.

    c) The voltage across R_2.

    d) Power consumed by R_1.

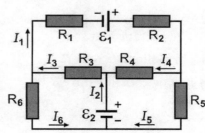

496 For the circuit shown below, the resistors are $R_1 = 7\,\Omega$, $R_2 = 4\,\Omega$, $R_3 = 6\,\Omega$, $R_4 = 7\,\Omega$, $R_5 = 10\,\Omega$ and $R_6 = 2\,\Omega$. The voltage sources are $\varepsilon_1 = 15$ V, $\varepsilon_2 = 7$ V and $\varepsilon_3 = 13$ V. Find:

    a) Currents.

    b) The voltage across R_1.

    c) The voltage across R_2.

    d) Power consumed by R_1.

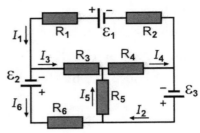

497 For the circuit shown below, the resistors are $R_1 = 3\,\Omega$, $R_2 = 10\,\Omega$, $R_3 = 4\,\Omega$, $R_4 = 6\,\Omega$, $R_5 = 12\,\Omega$ and $R_6 = 3\,\Omega$. The voltage sources are $\varepsilon_1 = 36$ V and $\varepsilon_2 = 16$ V. Find:

    a) Currents.

    b) The voltage across R_1.

    c) The voltage across R_2.

    d) Power consumed by R_1.

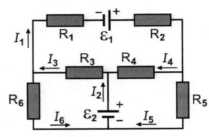

498 For the circuit shown below, the resistors are $R_1 = 6\,\Omega$, $R_2 = 11\,\Omega$, $R_3 = 11\,\Omega$, $R_4 = 2\,\Omega$, $R_5 = 6\,\Omega$ and $R_6 = 10\,\Omega$. The voltage sources are $\varepsilon_1 = 42$ V, $\varepsilon_2 = 8$ V and $\varepsilon_3 = 14$ V. Find:

    a) Currents.

    b) The voltage across R_1.

    c) The voltage across R_2.

    d) Power consumed by R_1.

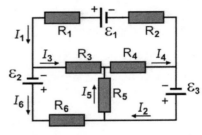

499 For the circuit shown below, the resistors are $R_1 = 6\,\Omega$, $R_2 = 4\,\Omega$, $R_3 = 1\,\Omega$, $R_4 = 3\,\Omega$, $R_5 = 12\,\Omega$ and $R_6 = 11\,\Omega$. The voltage sources are $\varepsilon_1 = 42$ V and $\varepsilon_2 = 9$ V. Find:

    a) Currents.

    b) The voltage across R_1.

    c) The voltage across R_2.

    d) Power consumed by R_1.

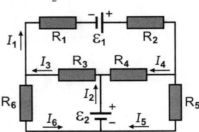

500 For the circuit shown below, the resistors are $R_1 = 2\,\Omega$, $R_2 = 7\,\Omega$, $R_3 = 6\,\Omega$, $R_4 = 9\,\Omega$, $R_5 = 6\,\Omega$ and $R_6 = 3\,\Omega$. The voltage sources are $\varepsilon_1 = 39$ V, $\varepsilon_2 = 13$ V and $\varepsilon_3 = 44$ V. Find:

    a) Currents.

    b) The voltage across R_1.

    c) The voltage across R_2.

    d) Power consumed by R_1.

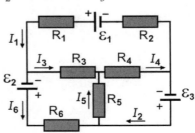

Solutions

1	126 s.
2	9.68 Ω.
3	0.466 Ω.
4	5.54×10^{-8} Ω·m.
5	0.0971 Ω.
6	**a)** 4.8 Ω, **b)** 4.656 V.
7	15.4 mC.
8	103 V.
9	470 μF.
10	210 V.
11	1.67 A.
12	70 Ω.
13	5.95 V.
14	66 C.
15	1.77 A.
16	536 s.
17	1.77 Ω.
18	174 Ω.
19	1.06×10^{-7} Ω·m.
20	12.7 Ω.
21	**a)** 5.1 Ω, **b)** 10.51 V.
22	279 μC.
23	70 V.
24	415 μF.
25	130 V.
26	1.14 A.
27	215 Ω.
28	5.84 V.
29	324 C.
30	47.6 mA.
31	53.7 s.
32	4.91 Ω.
33	1.23 Ω.
34	5.65×10^{-8} Ω·m.
35	5.54 Ω.
36	**a)** 2.9 Ω, **b)** 8.88 V.
37	7.04 mC.
38	19 V.
39	400 μF.
40	110 V.
41	1.24 A.
42	75 Ω.
43	3.66 V.
44	102 C.

45	850 mA.
46	108 s.
47	27.5 Ω.
48	0.445 Ω.
49	1.74×10^{-8} $\Omega \cdot$m.
50	21.1 Ω.
51	**a)** 3.7 Ω, **b)** 19.16 V.
52	10.26 mC.
53	44 V.
54	205 μF.
55	240 V.
56	0.889 A.
57	170 Ω.
58	21.3 V.
59	222 C.
60	1.37 A.
61	1130 s.
62	7.91 Ω.
63	0.456 Ω.
64	1.20×10^{-7} $\Omega \cdot$m.
65	0.205 Ω.
66	**a)** 4.8 Ω, **b)** 4.31 V.
67	1.98 mC.
68	8 V.
69	175 μF.
70	170 V.
71	0.627 A.
72	265 Ω.
73	10.8 V.
74	19100 C.
75	1.7 A.
76	**a)** 29.24 Ω, **b)** 0.8214 A, **c)** 18.07 V.
77	**a)** 60 Ω, **b)** 0.6333 A, **c)** 8.233 V.
78	**a)** 3.969 Ω, **b)** 2.444 A, **c)** 44 V.
79	**a)** 8.919 Ω, **b)** 1.267 A, **c)** 15.2 V.
80	**a)** 16.61 Ω, **b)** 1.577 A, **c)** 23.65 V.
81	**a)** 10.54 Ω, **b)** 0.2571 A, **c)** 6.171 V.
82	**a)** 26 Ω, **b)** 1.231 A, **c)** 24.62 V.
83	**a)** 4.286 Ω, **b)** 7.667 A, **c)** 46 V.
84	**a)** 13.42 Ω, **b)** 1.153 A, **c)** 13.84 V.
85	**a)** 64 Ω, **b)** 0.625 A, **c)** 18.13 V.
86	**a)** 9.442 Ω, **b)** 1.464 A, **c)** 41 V.
87	**a)** 14.19 Ω, **b)** 0.55 A, **c)** 10.45 V.
88	**a)** 16.07 Ω, **b)** 0.24 A, **c)** 5.04 V.
89	**a)** 12 Ω, **b)** 0.3214 A, **c)** 9 V.

90	**a)** 22 Ω, **b)** 2.136 A, **c)** 21.36 V.
91	**a)** 7.297 Ω, **b)** 3.8 A, **c)** 38 V.
92	**a)** 21.14 Ω, **b)** 0.3426 A, **c)** 8.222 V.
93	**a)** 50 Ω, **b)** 0.98 A, **c)** 21.56 V.
94	**a)** 3.224 Ω, **b)** 4.333 A, **c)** 26 V.
95	**a)** 8.049 Ω, **b)** 0.5333 A, **c)** 2.667 V.
96	**a)** 20.59 Ω, **b)** 0.6 A, **c)** 12 V.
97	**a)** 6.055 Ω, **b)** 0.9245 A, **c)** 24.04 V.
98	**a)** 38 Ω, **b)** 1.316 A, **c)** 23.68 V.
99	**a)** 3.059 Ω, **b)** 1.385 A, **c)** 18 V.
100	**a)** 28.31 Ω, **b)** 0.5868 A, **c)** 13.5 V.
101	**a)** 71 Ω, **b)** 0.4366 A, **c)** 6.986 V.
102	**a)** 7.293 Ω, **b)** 1.448 A, **c)** 42 V.
103	**a)** 13.43 Ω, **b)** 0.3448 A, **c)** 8.276 V.
104	**a)** 16.28 Ω, **b)** 1.407 A, **c)** 8.444 V.
105	**a)** 11.56 Ω, **b)** 0.425 A, **c)** 10.63 V.
106	**a)** 37 Ω, **b)** 0.973 A, **c)** 26.27 V.
107	**a)** 4.138 Ω, **b)** 8.4 A, **c)** 42 V.
108	**a)** 36.22 Ω, **b)** 0.4745 A, **c)** 9.965 V.
109	**a)** 69 Ω, **b)** 0.5652 A, **c)** 16.39 V.
110	**a)** 1.622 Ω, **b)** 1.5 A, **c)** 30 V.
111	**a)** 4.468 Ω, **b)** 0.8571 A, **c)** 22.29 V.
112	**a)** 14.98 Ω, **b)** 0.4783 A, **c)** 9.087 V.
113	**a)** 10.42 Ω, **b)** 0.1795 A, **c)** 5.026 V.
114	**a)** 25 Ω, **b)** 0.56 A, **c)** 10.08 V.
115	**a)** 11.48 Ω, **b)** 1.227 A, **c)** 27 V.
116	**a)** 9.632 Ω, **b)** 2.464 A, **c)** 19.72 V.
117	**a)** 69 Ω, **b)** 0.5362 A, **c)** 7.507 V.
118	**a)** 1.355 Ω, **b)** 11.67 A, **c)** 35 V.
119	**a)** 10.24 Ω, **b)** 1.714 A, **c)** 3.429 V.
120	**a)** 15.48 Ω, **b)** 0.5313 A, **c)** 11.16 V.
121	**a)** 1.7 Ω, **b)** 0.9744 A, **c)** 8.769 V.
122	**a)** 51 Ω, **b)** 0.2157 A, **c)** 4.745 V.
123	**a)** 4.105 Ω, **b)** 5.333 A, **c)** 32 V.
124	**a)** 31.44 Ω, **b)** 0.503 A, **c)** 2.515 V.
125	**a)** 34 Ω, **b)** 0.4412 A, **c)** 6.618 V.
126	**a)** 6.856 Ω, **b)** 2 A, **c)** 34 V.
127	**a)** 9.283 Ω, **b)** 0.2439 A, **c)** 7.317 V.
128	**a)** 24.44 Ω, **b)** 0.8364 A, **c)** 24.25 V.
129	**a)** 2.574 Ω, **b)** 0.6364 A, **c)** 1.909 V.
130	**a)** 20 Ω, **b)** 1.45 A, **c)** 20.3 V.
131	**a)** 8.919 Ω, **b)** 1.273 A, **c)** 28 V.
132	**a)** 37.77 Ω, **b)** 0.6393 A, **c)** 10.87 V.
133	**a)** 67 Ω, **b)** 0.3582 A, **c)** 10.03 V.
134	**a)** 3.857 Ω, **b)** 0.84 A, **c)** 21 V.

135   **a)** 12.35 Ω,   **b)** 1 A,   **c)** 5 V.
136   **a)** 9.221 Ω,   **b)** 0.5965 A,   **c)** 16.7 V.
137   **a)** 8.306 Ω,   **b)** 1.923 A,   **c)** 3.846 V.
138   **a)** 38 Ω,   **b)** 0.6579 A,   **c)** 15.13 V.
139   **a)** 6.107 Ω,   **b)** 1.526 A,   **c)** 29 V.
140   **a)** 31 Ω,   **b)** 0.3656 A,   **c)** 8.774 V.
141   **a)** 54 Ω,   **b)** 0.1667 A,   **c)** 3 V.
142   **a)** 7.13 Ω,   **b)** 1.524 A,   **c)** 32 V.
143   **a)** 13.89 Ω,   **b)** 0.561 A,   **c)** 6.732 V.
144   **a)** 16.75 Ω,   **b)** 0.8788 A,   **c)** 15.82 V.
145   **a)** 9.292 Ω,   **b)** 0.62 A,   **c)** 15.5 V.
146   **a)** 31 Ω,   **b)** 0.7419 A,   **c)** 5.935 V.
147   **a)** 6.618 Ω,   **b)** 0.88 A,   **c)** 22 V.
148   **a)** 39.98 Ω,   **b)** 0.41 A,   **c)** 11.89 V.
149   **a)** 33 Ω,   **b)** 1.455 A,   **c)** 21.82 V.
150   **a)** 5.859 Ω,   **b)** 2 A,   **c)** 32 V.
151   **a)** 3.5 µF,   **b)** 18.75 µC,   **c)** 3.75 V.
152   **a)** 4.846 µF,   **b)** 208.4 µC,   **c)** 23.15 V.
153   **a)** 56 µF,   **b)** 672 µC,   **c)** 24 V.
154   **a)** 36.33 µF,   **b)** 106.7 µC,   **c)** 3.556 V.
155   **a)** 15.53 µF,   **b)** 38.57 µC,   **c)** 2.143 V.
156   **a)** 24.05 µF,   **b)** 22.4 µC,   **c)** 0.8 V.
157   **a)** 12.24 µF,   **b)** 110.2 µC,   **c)** 4.408 V.
158   **a)** 25 µF,   **b)** 156 µC,   **c)** 13 V.
159   **a)** 13.36 µF,   **b)** 70.71 µC,   **c)** 7.857 V.
160   **a)** 7.63 µF,   **b)** 61.04 µC,   **c)** 2.348 V.
161   **a)** 48 µF,   **b)** 672 µC,   **c)** 32 V.
162   **a)** 21.28 µF,   **b)** 154 µC,   **c)** 22 V.
163   **a)** 10.77 µF,   **b)** 27.69 µC,   **c)** 3.077 V.
164   **a)** 13.59 µF,   **b)** 29.45 µC,   **c)** 14.73 V.
165   **a)** 2.1 µF,   **b)** 37.8 µC,   **c)** 12.6 V.
166   **a)** 54 µF,   **b)** 1276 µC,   **c)** 44 V.
167   **a)** 10.91 µF,   **b)** 290.7 µC,   **c)** 11.18 V.
168   **a)** 4.054 µF,   **b)** 166.2 µC,   **c)** 27.7 V.
169   **a)** 38 µF,   **b)** 126 µC,   **c)** 21 V.
170   **a)** 11.21 µF,   **b)** 303.8 µC,   **c)** 25.32 V.
171   **a)** 14.98 µF,   **b)** 195.8 µC,   **c)** 21.76 V.
172   **a)** 23.03 µF,   **b)** 120 µC,   **c)** 24 V.
173   **a)** 4.105 µF,   **b)** 98.53 µC,   **c)** 7.579 V.
174   **a)** 29 µF,   **b)** 264 µC,   **c)** 44 V.
175   **a)** 2.679 µF,   **b)** 20.25 µC,   **c)** 2.25 V.
176   **a)** 3.448 µF,   **b)** 82.76 µC,   **c)** 16.55 V.
177   **a)** 31 µF,   **b)** 330 µC,   **c)** 30 V.
178   **a)** 24.64 µF,   **b)** 366.4 µC,   **c)** 13.57 V.
179   **a)** 15.48 µF,   **b)** 87.23 µC,   **c)** 12.46 V.

180	**a)** 21.24 µF,	**b)** 124 µC,	**c)** 20.67 V.
181	**a)** 3.059 µF,	**b)** 113.2 µC,	**c)** 28.29 V.
182	**a)** 40 µF,	**b)** 133 µC,	**c)** 7 V.
183	**a)** 4.167 µF,	**b)** 120 µC,	**c)** 6 V.
184	**a)** 3.463 µF,	**b)** 117.8 µC,	**c)** 13.08 V.
185	**a)** 27 µF,	**b)** 598 µC,	**c)** 46 V.
186	**a)** 31.5 µF,	**b)** 98 µC,	**c)** 24.5 V.
187	**a)** 7.171 µF,	**b)** 41.25 µC,	**c)** 8.25 V.
188	**a)** 37.27 µF,	**b)** 258.8 µC,	**c)** 9.243 V.
189	**a)** 9.905 µF,	**b)** 356.6 µC,	**c)** 22.29 V.
190	**a)** 29 µF,	**b)** 432 µC,	**c)** 27 V.
191	**a)** 2.357 µF,	**b)** 19.93 µC,	**c)** 6.643 V.
192	**a)** 3.5 µF,	**b)** 133 µC,	**c)** 4.75 V.
193	**a)** 63 µF,	**b)** 168 µC,	**c)** 14 V.
194	**a)** 20.5 µF,	**b)** 275 µC,	**c)** 25 V.
195	**a)** 19.22 µF,	**b)** 123.4 µC,	**c)** 4.745 V.
196	**a)** 40.4 µF,	**b)** 126.3 µC,	**c)** 8.421 V.
197	**a)** 2.25 µF,	**b)** 54 µC,	**c)** 6 V.
198	**a)** 20 µF,	**b)** 152 µC,	**c)** 38 V.
199	**a)** 2.769 µF,	**b)** 21.54 µC,	**c)** 1.538 V.
200	**a)** 5.425 µF,	**b)** 92.22 µC,	**c)** 7.685 V.
201	**a)** 40 µF,	**b)** 962 µC,	**c)** 37 V.
202	**a)** 19.2 µF,	**b)** 56 µC,	**c)** 2.667 V.
203	**a)** 5.169 µF,	**b)** 79.2 µC,	**c)** 26.4 V.
204	**a)** 14.26 µF,	**b)** 123.2 µC,	**c)** 12.32 V.
205	**a)** 6.75 µF,	**b)** 108 µC,	**c)** 12 V.
206	**a)** 27 µF,	**b)** 330 µC,	**c)** 22 V.
207	**a)** 9.886 µF,	**b)** 221.5 µC,	**c)** 27.68 V.
208	**a)** 2.787 µF,	**b)** 52.96 µC,	**c)** 4.814 V.
209	**a)** 20 µF,	**b)** 161 µC,	**c)** 23 V.
210	**a)** 37.17 µF,	**b)** 208 µC,	**c)** 16 V.
211	**a)** 15.78 µF,	**b)** 70.21 µC,	**c)** 2.809 V.
212	**a)** 26.92 µF,	**b)** 44.88 µC,	**c)** 2.04 V.
213	**a)** 6.387 µF,	**b)** 261.9 µC,	**c)** 29.1 V.
214	**a)** 24 µF,	**b)** 104 µC,	**c)** 8 V.
215	**a)** 13.45 µF,	**b)** 173.6 µC,	**c)** 6.431 V.
216	**a)** 2.463 µF,	**b)** 14.78 µC,	**c)** 0.5277 V.
217	**a)** 36 µF,	**b)** 833 µC,	**c)** 49 V.
218	**a)** 32.67 µF,	**b)** 79.33 µC,	**c)** 3.778 V.
219	**a)** 18.95 µF,	**b)** 353.1 µC,	**c)** 29.42 V.
220	**a)** 20.12 µF,	**b)** 276.6 µC,	**c)** 18.44 V.
221	**a)** 9.767 µF,	**b)** 439.5 µC,	**c)** 15.7 V.
222	**a)** 39 µF,	**b)** 252 µC,	**c)** 21 V.
223	**a)** 8.974 µF,	**b)** 43.44 µC,	**c)** 3.949 V.
224	**a)** 6.491 µF,	**b)** 318.1 µC,	**c)** 15.15 V.

225	**a)** 59 μF, **b)** 529 μC, **c)** 23 V.
226	**a)** 1.23 A, **b)** 149 W.
227	8 V.
228	22.8 V, 7.85 Ω.
229	10.9 A, 50.4 Ω, 4.32×10^7 J = 12 kW·h.
230	440 mA , 465 Ω.
231	175 W, 7 Ω.
232	$4.25.
233	2.654×10^7 J, 7.371 kW·h.
234	**a)** 686.2 V, **b)** 531.9 mA.
235	**a)** 5.67 V, 13 V, 19.3 V, **b)** 3.22 W, 7.4 W, 10.9 W.
236	**a)** 2.13 A, **b)** 406 W.
237	18.3 V.
238	95.2 V, 113 Ω.
239	52.2 A, 11 Ω, 1.53×10^8 J = 42.5 kW·h.
240	1.8 A , 28.4 Ω.
241	234.9 W, 3.58 Ω.
242	$0.3.
243	2.192×10^7 J, 6.09 kW·h.
244	**a)** 639.7 V, **b)** 343.9 mA.
245	**a)** 4.67 V, 6.54 V, 7.79 V, **b)** 1.46 W, 2.04 W, 2.43 W.
246	**a)** 2.49 A, **b)** 566 W.
247	5.48 V.
248	8.89 V, 3.29 Ω.
249	50 A, 7.6 Ω, 1.25×10^8 J = 34.8 kW·h.
250	430 mA , 162 Ω.
251	125 W, 5 Ω.
252	$0.3.
253	3.525×10^7 J, 9.792 kW·h.
254	**a)** 392.3 V, **b)** 344.1 mA.
255	**a)** 2.04 V, 2.65 V, 3.31 V, **b)** 0.0964 W, 0.125 W, 0.157 W.
256	**a)** 2.31 A, **b)** 338 W.
257	16.5 V.
258	13.5 V, 2.75 Ω.
259	59.4 A, 8.5 Ω, 2.70×10^8 J = 75 kW·h.
260	4.5 A , 3.26 Ω.
261	152.5 W, 4.1 Ω.
262	$3.73.
263	2.883×10^7 J, 8.008 kW·h.
264	**a)** 110.2 V, **b)** 408.2 mA.
265	**a)** 9.95 V, 11.8 V, 14.2 V, **b)** 4.71 W, 5.61 W, 6.73 W.
266	**a)** 2.31 A, **b)** 415 W.
267	5.51 V.
268	267 V, 2220 Ω.
269	103 A, 3.6 Ω, 1.25×10^8 J = 34.8 kW·h.

270	2.6 A , 2.07 Ω.
271	257.4 W, 4.23 Ω.
272	$0.25.
273	1.270×10^7 J, 3.528 kW·h.
274	**a)** 562.8 V, **b)** 1.173 A.
275	**a)** 7.56 V, 11.1 V, 17.3 V, **b)** 3.36 W, 4.94 W, 7.7 W.
276	**a)** 3.91 A, **b)** 802 W.
277	35 V.
278	21.4 V, 5.94 Ω.
279	67.2 A, 8.63 Ω, 1.17×10^8 J = 32.5 kW·h.
280	370 mA , 131 Ω.
281	173.9 W, 7.87 Ω.
282	$0.34.
283	1.763×10^7 J, 4.896 kW·h.
284	**a)** 1335 V, **b)** 737.7 mA.
285	**a)** 8.21 V, 12.6 V, 15.2 V, **b)** 2.59 W, 3.99 W, 4.79 W.
286	**a)** 2.3 A, **b)** 426 W.
287	11.6 V.
288	12.4 V, 4.28 Ω.
289	6.2 A, 104 Ω, 1.08×10^7 J = 3 kW·h.
290	900 mA , 109 Ω.
291	112 W, 10.9 Ω.
292	$2.4.
293	6.047×10^7 J, 16.8 kW·h.
294	**a)** 707.2 V, **b)** 579.7 mA.
295	**a)** 6.78 V, 8.82 V, 10.4 V, **b)** 1.07 W, 1.39 W, 1.64 W.
296	**a)** 286 nF, **b)** 11.21 mJ.
297	90 V, 3.447 mJ.
298	**a)** 261 nF, **b)** 46.98 μC.
299	**a)** 34.48 J, **b)** 218.9 mC.
300	**a)** 535 V, **b)** 11.23 μC.
301	**a)** 75.4 mC, **b)** 19.6 J.
302	**a)** 220 V, **b)** 456 nF.
303	**a)** 445 μF, **b)** 24.97 J.
304	20 V, 124 mJ.
305	**a)** 541 nF, **b)** 162.3 μC.
306	**a)** 732.1 μJ, **b)** 13.31 μC.
307	**a)** 200 V, **b)** 44 mC.
308	**a)** 22.04 μC, **b)** 3.196 mJ.
309	**a)** 230 V, **b)** 921 nF.
310	**a)** 195 μF, **b)** 4.507 J.
311	245 V, 21.34 mJ.
312	**a)** 371 nF, **b)** 133.6 μC.
313	**a)** 242.5 mJ, **b)** 642.5 μC.
314	**a)** 540 V, **b)** 95.04 μC.

315	**a)** 2.25 µC, **b)** 421.9 µJ.
316	**a)** 485 V, **b)** 651 nF.
317	**a)** 10 µF, **b)** 3.916 J.
318	415 V, 4.392 mJ.
319	**a)** 331 nF, **b)** 31.45 µC.
320	**a)** 216 mJ, **b)** 7.2 mC.
321	**a)** 165 V, **b)** 103.3 µC.
322	**a)** 92.8 mC, **b)** 26.91 J.
323	**a)** 650 V, **b)** 270 µF.
324	**a)** 780 µF, **b)** 60.85 J.
325	235 V, 6.793 mJ.
326	**a)** 191 nF, **b)** 91.68 µC.
327	**a)** 247.3 J, **b)** 646.4 mC.
328	**a)** 850 V, **b)** 514.3 mC.
329	**a)** 569.9 mC, **b)** 218 J.
330	**a)** 640 V, **b)** 395 µF.
331	**a)** 370 µF, **b)** 78.16 J.
332	605 V, 114.6 mJ.
333	**a)** 531 nF, **b)** 464.6 µC.
334	**a)** 225 mJ, **b)** 6 mC.
335	**a)** 725 V, **b)** 174 mC.
336	**a)** 224.9 µC, **b)** 82.65 mJ.
337	**a)** 285 V, **b)** 290 µF.
338	**a)** 615 µF, **b)** 214.4 J.
339	530 V, 702.2 mJ.
340	**a)** 265 µF, **b)** 152.4 mC.
341	**a)** 54.88 mJ, **b)** 175.6 µC.
342	**a)** 415 V, **b)** 243.2 µC.
343	**a)** 103.1 µC, **b)** 12.64 mJ.
344	**a)** 360 V, **b)** 81 nF.
345	**a)** 561 nF, **b)** 159.9 mJ.
346	**a)** $I = 2.471$ A
	b) 12.35 V, **c)** 7.412 V, **d)** 30.52 W
347	**a)** $I = 0.6154$ A
	b) 3.077 V, **c)** 3.077 V, **d)** 1.893 W
348	**a)** $I = 2.941$ A
	b) 14.71 V, **c)** 20.59 V, **d)** 43.25 W
349	**a)** $I = 0.7$ A
	b) 5.6 V, **c)** 1.4 V, **d)** 3.92 W
350	**a)** $I = 2.435$ A
	b) 26.78 V, **c)** 17.04 V, **d)** 65.21 W
351	**a)** $I = 2$ A
	b) 10 V, **c)** 4 V, **d)** 20 W
352	**a)** $I = 1.478$ A
	b) 16.26 V, **c)** 11.83 V, **d)** 24.04 W

353 **a)** $I = 0.375$ A
b) 0.75 V,   **c)** 1.875 V,   **d)** 0.2813 W

354 **a)** $I = 2.188$ A
b) 8.75 V,   **c)** 15.31 V,   **d)** 19.14 W

355 **a)** $I = 0.8333$ A
b) 4.167 V,   **c)** 5.833 V,   **d)** 3.472 W

356 **a)** $I = 1.818$ A
b) 20 V,   **c)** 16.36 V,   **d)** 36.36 W

357 **a)** $I = 0.9375$ A
b) 2.813 V,   **c)** 9.375 V,   **d)** 2.637 W

358 **a)** $I = 3.278$ A
b) 26.22 V,   **c)** 13.11 V,   **d)** 85.95 W

359 **a)** $I = 0.9032$ A
b) 9.032 V,   **c)** 10.84 V,   **d)** 8.158 W

360 **a)** $I = 1.2$ A
b) 7.2 V,   **c)** 7.2 V,   **d)** 8.64 W

361 **a)** $I = 1.846$ A
b) 5.538 V,   **c)** 9.231 V,   **d)** 10.22 W

362 **a)** $I = 4.313$ A
b) 12.94 V,   **c)** 25.88 V,   **d)** 55.79 W

363 **a)** $I = 0.6154$ A
b) 1.846 V,   **c)** 7.385 V,   **d)** 1.136 W

364 **a)** $I = 2.643$ A
b) 18.5 V,   **c)** 5.286 V,   **d)** 48.89 W

365 **a)** $I = 0.7692$ A
b) 3.077 V,   **c)** 7.692 V,   **d)** 2.367 W

366 **a)** $I = 3.034$ A
b) 27.31 V,   **c)** 27.31 V,   **d)** 82.87 W

367 **a)** $I = 0.1667$ A
b) 0.5 V,   **c)** 0.8333 V,   **d)** 0.08333 W

368 **a)** $I = 2.786$ A
b) 25.07 V,   **c)** 5.571 V,   **d)** 69.84 W

369 **a)** $I = 0.3889$ A
b) 4.278 V,   **c)** 2.333 V,   **d)** 1.664 W

370 **a)** $I = 0.9524$ A
b) 10.48 V,   **c)** 2.857 V,   **d)** 9.977 W

371 **a)** $I = 1.421$ A
b) 15.63 V,   **c)** 9.947 V,   **d)** 22.21 W

372 **a)** $I = 1.742$ A
b) 20.9 V,   **c)** 12.19 V,   **d)** 36.41 W

373 **a)** $I = 0.2857$ A
b) 1.429 V,   **c)** 2.857 V,   **d)** 0.4082 W

374 **a)** $I = 2.091$ A
b) 23 V,   **c)** 16.73 V,   **d)** 48.09 W

375   **a)** $I = 0.5$ A

        b) 3 V,   **c)** 2.5 V,   **d)** 1.5 W

376   **a)** $I = 4.429$ A

        b) 13.29 V,   **c)** 22.14 V,   **d)** 58.84 W

377   **a)** $I = 1.636$ A

        b) 4.909 V,   **c)** 6.545 V,   **d)** 8.033 W

378   **a)** $I = 0.9231$ A

        b) 6.462 V,   **c)** 6.462 V,   **d)** 5.964 W

379   **a)** $I = 2.364$ A

        b) 9.455 V,   **c)** 11.82 V,   **d)** 22.35 W

380   **a)** $I = 2.071$ A

        b) 8.286 V,   **c)** 24.86 V,   **d)** 17.16 W

381   **a)** $I = 0.625$ A

        b) 5.625 V,   **c)** 6.875 V,   **d)** 3.516 W

382   **a)** $I = 5$ A

        b) 40 V,   **c)** 10 V,   **d)** 200 W

383   **a)** $I = 0.5$ A

        b) 3 V,   **c)** 2.5 V,   **d)** 1.5 W

384   **a)** $I = 2.286$ A

        b) 4.571 V,   **c)** 18.29 V,   **d)** 10.45 W

385   **a)** $I = 0.1538$ A

        b) 1.692 V,   **c)** 0.6154 V,   **d)** 0.2604 W

386   **a)** $I = 1.85$ A

        b) 14.8 V,   **c)** 12.95 V,   **d)** 27.38 W

387   **a)** $I = 1.286$ A

        b) 3.857 V,   **c)** 5.143 V,   **d)** 4.959 W

388   **a)** $I = 1.467$ A

        b) 2.933 V,   **c)** 8.8 V,   **d)** 4.302 W

389   **a)** $I = 2$ A

        b) 8 V,   **c)** 8 V,   **d)** 16 W

390   **a)** $I = 1.2$ A

        b) 14.4 V,   **c)** 9.6 V,   **d)** 17.28 W

391   **a)** $I = 1.524$ A

        b) 6.095 V,   **c)** 15.24 V,   **d)** 9.288 W

392   **a)** $I = 1.111$ A

        b) 6.667 V,   **c)** 4.444 V,   **d)** 7.407 W

393   **a)** $I = 0.1154$ A

        b) 1.269 V,   **c)** 0.6923 V,   **d)** 0.1464 W

394   **a)** $I = 2.81$ A

        b) 28.1 V,   **c)** 8.429 V,   **d)** 78.93 W

395   **a)** $I = 1.037$ A

        b) 11.41 V,   **c)** 7.259 V,   **d)** 11.83 W

396   **a)** $I_1 = 3.415$ A, $I_2 = 1.894$ A, $I_3 = 1.52$ A

        b) 27.32 V,   **c)** 5.683 V,   **d)** 93.28 W

397   **a)** $I_1 = 3.68$ A, $I_2 = 3.04$ A, $I_3 = 0.64$ A

       b) 14.72 V,   **c)** 21.28 V,   **d)** 54.17 W

398   **a)** $I_1 = 1.464$ A, $I_2 = 0.7536$ A, $I_3 = 2.217$ A

       b) 8.783 V,   **c)** 6.783 V,   **d)** 12.86 W

399   **a)** $I_1 = 4.679$ A, $I_2 = 3.299$ A, $I_3 = 1.381$ A

       b) 18.72 V,   **c)** 19.79 V,   **d)** 87.58 W

400   **a)** $I_1 = 1.156$ A, $I_2 = 0.4573$ A, $I_3 = 0.6985$ A

       b) 2.312 V,   **c)** 1.372 V,   **d)** 2.672 W

401   **a)** $I_1 = 2.091$ A, $I_2 = 0.7273$ A, $I_3 = 1.364$ A

       b) 23 V,   **c)** 8 V,   **d)** 48.09 W

402   **a)** $I_1 = 2.821$ A, $I_2 = 1.513$ A, $I_3 = 1.308$ A

       b) 25.38 V,   **c)** 13.62 V,   **d)** 71.6 W

403   **a)** $I_1 = 0.5676$ A, $I_2 = 2.473$ A, $I_3 = 3.041$ A

       b) 2.838 V,   **c)** 14.84 V,   **d)** 1.611 W

404   **a)** $I_1 = 5.618$ A, $I_2 = 4.329$ A, $I_3 = 1.289$ A

       b) 33.71 V,   **c)** 38.96 V,   **d)** 189.4 W

405   **a)** $I_1 = 0.6221$ A, $I_2 = 0.2714$ A, $I_3 = 0.3507$ A

       b) 1.866 V,   **c)** 1.9 V,   **d)** 1.161 W

406   **a)** $I_1 = 3.987$ A, $I_2 = 2.354$ A, $I_3 = 1.633$ A

       b) 19.94 V,   **c)** 7.063 V,   **d)** 79.49 W

407   **a)** $I_1 = 2.75$ A, $I_2 = 1.75$ A, $I_3 = 1$ A

       b) 11 V,   **c)** 14 V,   **d)** 30.25 W

408   **a)** $I_1 = 1.571$ A, $I_2 = 0.4643$ A, $I_3 = 2.036$ A

       b) 12.57 V,   **c)** 5.571 V,   **d)** 19.76 W

409   **a)** $I_1 = 2.41$ A, $I_2 = 1.593$ A, $I_3 = 0.8166$ A

       b) 28.92 V,   **c)** 17.52 V,   **d)** 69.68 W

410   **a)** $I_1 = 2.887$ A, $I_2 = 0.434$ A, $I_3 = 2.453$ A

       b) 11.55 V,   **c)** 3.038 V,   **d)** 33.33 W

411   **a)** $I_1 = 1.865$ A, $I_2 = 0.6346$ A, $I_3 = 1.231$ A

       b) 14.92 V,   **c)** 5.077 V,   **d)** 27.84 W

412   **a)** $I_1 = 2.704$ A, $I_2 = 2.58$ A, $I_3 = 0.1235$ A

       b) 13.52 V,   **c)** 15.48 V,   **d)** 36.55 W

413   **a)** $I_1 = 0.4375$ A, $I_2 = 0.5469$ A, $I_3 = 0.9844$ A

       b) 2.188 V,   **c)** 2.188 V,   **d)** 0.957 W

414   **a)** $I_1 = 2.228$ A, $I_2 = 1.916$ A, $I_3 = 0.3119$ A

       b) 24.5 V,   **c)** 5.748 V,   **d)** 54.59 W

415   **a)** $I_1 = 1.646$ A, $I_2 = 0.3248$ A, $I_3 = 1.321$ A

       b) 14.81 V,   **c)** 2.274 V,   **d)** 24.38 W

416   **a)** $I_1 = 3.576$ A, $I_2 = 0.8898$ A, $I_3 = 2.686$ A

       b) 17.88 V,   **c)** 7.119 V,   **d)** 63.95 W

417   **a)** $I_1 = 3.179$ A, $I_2 = 1.214$ A, $I_3 = 1.964$ A

       b) 19.07 V,   **c)** 10.93 V,   **d)** 60.62 W

418   **a)** $I_1 = 0.241$ A, $I_2 = 2.157$ A, $I_3 = 2.398$ A

       b) 2.41 V,   **c)** 19.41 V,   **d)** 0.5806 W

419 **a)** $I_1 = 7.609$ A, $I_2 = 2.826$ A, $I_3 = 4.783$ A
b) 30.43 V, **c)** 11.3 V, **d)** 231.6 W

420 **a)** $I_1 = 3.011$ A, $I_2 = 0.8533$ A, $I_3 = 2.158$ A
b) 9.033 V, **c)** 9.386 V, **d)** 27.2 W

421 **a)** $I_1 = 5.542$ A, $I_2 = 0.8333$ A, $I_3 = 4.708$ A
b) 22.17 V, **c)** 5.833 V, **d)** 122.8 W

422 **a)** $I_1 = 3.506$ A, $I_2 = 2.696$ A, $I_3 = 0.8101$ A
b) 10.52 V, **c)** 13.48 V, **d)** 36.88 W

423 **a)** $I_1 = 1.795$ A, $I_2 = 0.1233$ A, $I_3 = 1.918$ A
b) 19.74 V, **c)** 0.7397 V, **d)** 35.42 W

424 **a)** $I_1 = 2.25$ A, $I_2 = 1.333$ A, $I_3 = 0.9167$ A
b) 18 V, **c)** 14.67 V, **d)** 40.5 W

425 **a)** $I_1 = 0.5474$ A, $I_2 = 0.3431$ A, $I_3 = 0.2044$ A
b) 5.474 V, **c)** 2.058 V, **d)** 2.997 W

426 **a)** $I_1 = 2.658$ A, $I_2 = 0.6913$ A, $I_3 = 1.966$ A
b) 29.23 V, **c)** 2.765 V, **d)** 77.7 W

427 **a)** $I_1 = 3.145$ A, $I_2 = 2.821$ A, $I_3 = 0.3241$ A
b) 9.434 V, **c)** 22.57 V, **d)** 29.67 W

428 **a)** $I_1 = 0.1$ A, $I_2 = 2.6$ A, $I_3 = 2.7$ A
b) 0.2 V, **c)** 18.2 V, **d)** 0.02 W

429 **a)** $I_1 = 2.955$ A, $I_2 = 1.773$ A, $I_3 = 1.182$ A
b) 23.64 V, **c)** 7.091 V, **d)** 69.83 W

430 **a)** $I_1 = 1.841$ A, $I_2 = 0.2698$ A, $I_3 = 1.571$ A
b) 5.524 V, **c)** 1.079 V, **d)** 10.17 W

431 **a)** $I_1 = 2.375$ A, $I_2 = 1.05$ A, $I_3 = 1.325$ A
b) 23.75 V, **c)** 5.25 V, **d)** 56.41 W

432 **a)** $I_1 = 2.452$ A, $I_2 = 2.133$ A, $I_3 = 0.3193$ A
b) 9.807 V, **c)** 19.19 V, **d)** 24.05 W

433 **a)** $I_1 = 0.2059$ A, $I_2 = 4.618$ A, $I_3 = 4.824$ A
b) 1.235 V, **c)** 9.235 V, **d)** 0.2543 W

434 **a)** $I_1 = 5.737$ A, $I_2 = 4.591$ A, $I_3 = 1.146$ A
b) 28.68 V, **c)** 27.54 V, **d)** 164.6 W

435 **a)** $I_1 = 1.771$ A, $I_2 = 0.1271$ A, $I_3 = 1.644$ A
b) 7.085 V, **c)** 1.525 V, **d)** 12.55 W

436 **a)** $I_1 = 2.79$ A, $I_2 = 0.5161$ A, $I_3 = 2.274$ A
b) 27.9 V, **c)** 3.097 V, **d)** 77.86 W

437 **a)** $I_1 = 2.159$ A, $I_2 = 1.794$ A, $I_3 = 0.3651$ A
b) 17.27 V, **c)** 19.73 V, **d)** 37.28 W

438 **a)** $I_1 = 3.13$ A, $I_2 = 0.9565$ A, $I_3 = 4.087$ A
b) 15.65 V, **c)** 7.652 V, **d)** 49 W

439 **a)** $I_1 = 2.626$ A, $I_2 = 1.289$ A, $I_3 = 1.337$ A
b) 23.63 V, **c)** 14.18 V, **d)** 62.06 W

440 **a)** $I_1 = 2.441$ A, $I_2 = 0.5315$ A, $I_3 = 1.91$ A
b) 7.324 V, **c)** 2.126 V, **d)** 17.88 W

441   **a)** $I_1 = 1.507$ A, $I_2 = 0.1778$ A, $I_3 = 1.33$ A

        b) 9.044 V,   **c)** 1.956 V,   **d)** 13.63 W

442   **a)** $I_1 = 3.5$ A, $I_2 = 3.333$ A, $I_3 = 0.1667$ A

        b) 14 V,   **c)** 20 V,   **d)** 49 W

443   **a)** $I_1 = 5.687$ A, $I_2 = 0.1928$ A, $I_3 = 5.88$ A

        b) 34.12 V,   **c)** 2.12 V,   **d)** 194 W

444   **a)** $I_1 = 3.571$ A, $I_2 = 2.429$ A, $I_3 = 1.143$ A

        b) 17.86 V,   **c)** 19.43 V,   **d)** 63.78 W

445   **a)** $I_1 = 0.5987$ A, $I_2 = 0.3676$ A, $I_3 = 0.2311$ A

        b) 4.191 V,   **c)** 2.941 V,   **d)** 2.509 W

446   **a)** $I_1 = 2.591$ A, $I_2 = 0.9773$ A, $I_3 = 1.614$ A

        b) 31.09 V,   **c)** 3.909 V,   **d)** 80.55 W

447   **a)** $I_1 = 2.497$ A, $I_2 = 1.774$ A, $I_3 = 0.7236$ A

        b) 12.49 V,   **c)** 19.51 V,   **d)** 31.19 W

448   **a)** $I_1 = 0.9515$ A, $I_2 = 0.2524$ A, $I_3 = 1.204$ A

        b) 4.757 V,   **c)** 0.7573 V,   **d)** 4.526 W

449   **a)** $I_1 = 4.29$ A, $I_2 = 2.562$ A, $I_3 = 1.728$ A

        b) 21.45 V,   **c)** 30.74 V,   **d)** 92.01 W

450   **a)** $I_1 = 1.613$ A, $I_2 = 1.338$ A, $I_3 = 0.2753$ A

        b) 16.13 V,   **c)** 2.676 V,   **d)** 26.03 W

451   **a)** $I_1 = 2.513$ A, $I_2 = 3.544$ A, $I_3 = 4.081$ A

        $I_4 = 0.5371$ A, $I_5 = 1.976$ A, $I_6 = 1.568$ A

        b) 7.539 V,   **c)** 7.539 V,   **d)** 18.95 W

452   **a)** $I_1 = 1.415$ A, $I_2 = 1.669$ A, $I_3 = 0.8769$ A

        $I_4 = 3.085$ A, $I_5 = 2.208$ A, $I_6 = 0.5385$ A

        b) 5.662 V,   **c)** 15.57 V,   **d)** 8.013 W

453   **a)** $I_1 = 1.052$ A, $I_2 = 2.869$ A, $I_3 = 3.017$ A

        $I_4 = 0.148$ A, $I_5 = 0.9036$ A, $I_6 = 1.966$ A

        b) 10.52 V,   **c)** 8.413 V,   **d)** 11.06 W

454   **a)** $I_1 = 1.088$ A, $I_2 = 1.166$ A, $I_3 = 0.5647$ A

        $I_4 = 2.255$ A, $I_5 = 1.69$ A, $I_6 = 0.5237$ A

        b) 3.265 V,   **c)** 11.97 V,   **d)** 3.554 W

455   **a)** $I_1 = 1.102$ A, $I_2 = 2.578$ A, $I_3 = 2.801$ A

        $I_4 = 0.2228$ A, $I_5 = 0.8789$ A, $I_6 = 1.699$ A

        b) 11.02 V,   **c)** 7.712 V,   **d)** 12.14 W

456   **a)** $I_1 = 0.8463$ A, $I_2 = 2.451$ A, $I_3 = 0.3209$ A

        $I_4 = 3.297$ A, $I_5 = 2.976$ A, $I_6 = 0.5253$ A

        b) 8.463 V,   **c)** 3.385 V,   **d)** 7.162 W

457   **a)** $I_1 = 2.403$ A, $I_2 = 4.749$ A, $I_3 = 5.161$ A

        $I_4 = 0.4125$ A, $I_5 = 1.991$ A, $I_6 = 2.758$ A

        b) 14.42 V,   **c)** 7.21 V,   **d)** 34.65 W

458 **a)** $I_1 = 0.6038$ A, $I_2 = 1.111$ A, $I_3 = 0.1747$ A
$I_4 = 1.714$ A, $I_5 = 1.54$ A, $I_6 = 0.4291$ A
b) 6.038 V, **c)** 7.246 V, **d)** 3.646 W

459 **a)** $I_1 = 2.287$ A, $I_2 = 2.997$ A, $I_3 = 3.179$ A
$I_4 = 0.1826$ A, $I_5 = 2.104$ A, $I_6 = 0.8924$ A
b) 11.44 V, **c)** 18.3 V, **d)** 26.15 W

460 **a)** $I_1 = 0.8103$ A, $I_2 = 2.879$ A, $I_3 = 0.1271$ A
$I_4 = 3.689$ A, $I_5 = 3.562$ A, $I_6 = 0.6832$ A
b) 7.293 V, **c)** 3.241 V, **d)** 5.91 W

461 **a)** $I_1 = 1.636$ A, $I_2 = 1.237$ A, $I_3 = 1.977$ A
$I_4 = 0.7401$ A, $I_5 = 0.896$ A, $I_6 = 0.341$ A
b) 11.45 V, **c)** 19.63 V, **d)** 18.74 W

462 **a)** $I_1 = 0.9595$ A, $I_2 = 0.551$ A, $I_3 = 0.4312$ A
$I_4 = 1.511$ A, $I_5 = 1.079$ A, $I_6 = 0.5283$ A
b) 3.838 V, **c)** 2.879 V, **d)** 3.683 W

463 **a)** $I_1 = 2.831$ A, $I_2 = 2.251$ A, $I_3 = 3.365$ A
$I_4 = 1.114$ A, $I_5 = 1.717$ A, $I_6 = 0.5338$ A
b) 11.32 V, **c)** 8.493 V, **d)** 32.05 W

464 **a)** $I_1 = 1.083$ A, $I_2 = 3.397$ A, $I_3 = 0.1983$ A
$I_4 = 4.479$ A, $I_5 = 4.281$ A, $I_6 = 0.8843$ A
b) 2.165 V, **c)** 5.413 V, **d)** 2.344 W

465 **a)** $I_1 = 0.9938$ A, $I_2 = 1.265$ A, $I_3 = 1.497$ A
$I_4 = 0.2316$ A, $I_5 = 0.7621$ A, $I_6 = 0.5031$ A
b) 9.938 V, **c)** 7.95 V, **d)** 9.876 W

466 **a)** $I_1 = 2.156$ A, $I_2 = 0.9522$ A, $I_3 = 1.011$ A
$I_4 = 3.108$ A, $I_5 = 2.097$ A, $I_6 = 1.144$ A
b) 12.93 V, **c)** 8.622 V, **d)** 27.88 W

467 **a)** $I_1 = 2.119$ A, $I_2 = 2.34$ A, $I_3 = 3.079$ A
$I_4 = 0.7394$ A, $I_5 = 1.38$ A, $I_6 = 0.9603$ A
b) 21.19 V, **c)** 8.476 V, **d)** 44.9 W

468 **a)** $I_1 = 0.9972$ A, $I_2 = 0.9606$ A, $I_3 = 0.4903$ A
$I_4 = 1.958$ A, $I_5 = 1.468$ A, $I_6 = 0.5069$ A
b) 8.975 V, **c)** 10.97 V, **d)** 8.95 W

469 **a)** $I_1 = 1.728$ A, $I_2 = 2.102$ A, $I_3 = 2.382$ A
$I_4 = 0.28$ A, $I_5 = 1.448$ A, $I_6 = 0.6544$ A
b) 12.1 V, **c)** 17.28 V, **d)** 20.9 W

470 **a)** $I_1 = 2.906$ A, $I_2 = 2.053$ A, $I_3 = 0.3422$ A
$I_4 = 4.959$ A, $I_5 = 4.617$ A, $I_6 = 2.563$ A
b) 8.717 V, **c)** 20.34 V, **d)** 25.33 W

471 **a)** $I_1 = 1.571$ A, $I_2 = 5.041$ A, $I_3 = 5.19$ A
$I_4 = 0.1489$ A, $I_5 = 1.422$ A, $I_6 = 3.62$ A
b) 15.71 V, **c)** 17.28 V, **d)** 24.67 W

472 **a)** $I_1 = 1.456$ A, $I_2 = 3.521$ A, $I_3 = 0.8244$ A

              $I_4 = 4.977$ A, $I_5 = 4.153$ A, $I_6 = 0.6317$ A

           b) 14.56 V,   **c)** 2.912 V,   **d)** 21.2 W

473 **a)** $I_1 = 2.307$ A, $I_2 = 2.38$ A, $I_3 = 3.031$ A

              $I_4 = 0.6506$ A, $I_5 = 1.656$ A, $I_6 = 0.7245$ A

           b) 13.84 V,   **c)** 9.226 V,   **d)** 31.92 W

474 **a)** $I_1 = 1.104$ A, $I_2 = 1.722$ A, $I_3 = 0.154$ A

              $I_4 = 2.826$ A, $I_5 = 2.672$ A, $I_6 = 0.9499$ A

           b) 6.623 V,   **c)** 7.727 V,   **d)** 7.312 W

475 **a)** $I_1 = 2.036$ A, $I_2 = 2.068$ A, $I_3 = 2.316$ A

              $I_4 = 0.2479$ A, $I_5 = 1.788$ A, $I_6 = 0.2806$ A

           b) 16.28 V,   **c)** 20.36 V,   **d)** 33.15 W

476 **a)** $I_1 = 2.11$ A, $I_2 = 1.087$ A, $I_3 = 0.8677$ A

              $I_4 = 3.197$ A, $I_5 = 2.329$ A, $I_6 = 1.242$ A

           b) 6.329 V,   **c)** 4.219 V,   **d)** 13.35 W

477 **a)** $I_1 = 1.953$ A, $I_2 = 4.391$ A, $I_3 = 4.984$ A

              $I_4 = 0.5938$ A, $I_5 = 1.359$ A, $I_6 = 3.031$ A

           b) 15.63 V,   **c)** 7.813 V,   **d)** 30.52 W

478 **a)** $I_1 = 1.498$ A, $I_2 = 1.032$ A, $I_3 = 0.7961$ A

              $I_4 = 2.531$ A, $I_5 = 1.735$ A, $I_6 = 0.7023$ A

           b) 13.49 V,   **c)** 7.492 V,   **d)** 20.21 W

479 **a)** $I_1 = 1.731$ A, $I_2 = 1.934$ A, $I_3 = 2.359$ A

              $I_4 = 0.4246$ A, $I_5 = 1.306$ A, $I_6 = 0.6282$ A

           b) 17.31 V,   **c)** 17.31 V,   **d)** 29.95 W

480 **a)** $I_1 = 1.153$ A, $I_2 = 1.35$ A, $I_3 = 0.2083$ A

              $I_4 = 2.503$ A, $I_5 = 2.294$ A, $I_6 = 0.9445$ A

           b) 3.458 V,   **c)** 5.764 V,   **d)** 3.987 W

481 **a)** $I_1 = 2.604$ A, $I_2 = 4.212$ A, $I_3 = 4.569$ A

              $I_4 = 0.3575$ A, $I_5 = 2.246$ A, $I_6 = 1.965$ A

           b) 10.42 V,   **c)** 18.23 V,   **d)** 27.12 W

482 **a)** $I_1 = 1.848$ A, $I_2 = 1.494$ A, $I_3 = 0.6056$ A

              $I_4 = 3.342$ A, $I_5 = 2.737$ A, $I_6 = 1.243$ A

           b) 5.545 V,   **c)** 9.241 V,   **d)** 10.25 W

483 **a)** $I_1 = 1.442$ A, $I_2 = 1.917$ A, $I_3 = 2.271$ A

              $I_4 = 0.3536$ A, $I_5 = 1.088$ A, $I_6 = 0.8287$ A

           b) 11.54 V,   **c)** 15.86 V,   **d)** 16.63 W

484 **a)** $I_1 = 2.311$ A, $I_2 = 1.762$ A, $I_3 = 0.8942$ A

              $I_4 = 4.072$ A, $I_5 = 3.178$ A, $I_6 = 1.417$ A

           b) 11.55 V,   **c)** 13.86 V,   **d)** 26.7 W

485 **a)** $I_1 = 2.133$ A, $I_2 = 2.434$ A, $I_3 = 2.705$ A

              $I_4 = 0.2712$ A, $I_5 = 1.862$ A, $I_6 = 0.5723$ A

           b) 12.8 V,   **c)** 21.33 V,   **d)** 27.29 W

Made in the USA
Monee, IL
31 October 2024